ENVIRONMENTAL REMEDIATION TECHNOLOGIES, REGULATIONS AND SAFETY

THE EUROPEAN UNION'S EMISSIONS TRADING SCHEME

CLIMATE AND AVIATION CONSIDERATIONS

ENVIRONMENTAL REMEDIATION TECHNOLOGIES, REGULATIONS AND SAFETY

Additional books in this series can be found on Nova's website under the Series tab.

Additional e-books in this series can be found on Nova's website under the e-books tab.

CLIMATE CHANGE AND ITS CAUSES, EFFECTS AND PREDICTION

Additional books in this series can be found on Nova's website under the Series tab.

Additional e-books in this series can be found on Nova's website under the e-books tab.

ENVIRONMENTAL REMEDIATION TECHNOLOGIES,
REGULATIONS AND SAFETY

THE EUROPEAN UNION'S EMISSIONS TRADING SCHEME

CLIMATE AND AVIATION CONSIDERATIONS

ANTHONY R. MOORE
AND
MIA F. RYAN
EDITORS

NOVINKA

New York

For permission to use material from this book please contact us:
Telephone 631-231-7269; Fax 631-231-8175
Web Site: http://www.novapublishers.com

NOTICE TO THE READER

LIBRARY OF CONGRESS CATALOGING-IN-PUBLICATION DATA

ISBN: 978-1-62081-955-5

Published by Nova Science Publishers, Inc. † New York

CONTENTS

PREFACE

Climate change is generally viewed as a global issue, but proposed responses typically require action at the national level. With the 1997 Kyoto Protocol now in force and setting emissions objectives for 2008-2012, countries that ratified the protocol are implementing strategies to begin reducing their emissions of greenhouse gases. The European Union's (EU) Emissions Trading Scheme (ETS) is a cornerstone of the EU's efforts to meet its obligation under the Kyoto Protocol. It covers more than 10,000 energy intensive facilities across the 27 EU Member countries; covered entities emit about 45% of the EU's carbon dioxide emissions. This book examines the European Union's Emissions Trading Scheme (ETS) with a focus on climate and aviation considerations.

Chapter 1 – The European Union's (EU) Emissions Trading Scheme (ETS) is a cornerstone of the EU's efforts to meet its obligation under the Kyoto Protocol. It covers more than 10,000 energy intensive facilities across the 27 EU Member countries; covered entities emit about 45% of the EU's carbon dioxide emissions. A "Phase 1" trading period began January 1, 2005. A second, Phase 2, trading period began in 2008, covering the period of the Kyoto Protocol. A Phase 3 will begin in 2013 designed to reduce emissions by 21% from 2005 levels.

Several positive results from the Phase 1 "learning by doing" exercise assisted the ETS in making the Phase 2 process run more smoothly, including: (1) greatly improving emissions data, (2) encouraging development of the Kyoto Protocol's project-based mechanisms—Clean Development Mechanism (CDM) and Joint Implementation (JI), and (3) influencing corporate behavior to begin pricing in the value of allowances in decision-making, particularly in the electric utility sector.

However, several issues that arose during the first phase were not resolved as the ETS implemented Phase 2, including allocation schemes and new entrant reserves, and others. A more comprehensive and coordinated response has been designed for Phase 3 with harmonized and coordinated rules being developed by the European Commission.

The United States is not a party to the Kyoto Protocol. However, six years of carbon emissions trading has given the EU valuable experience in designing and operating a greenhouse gas trading system. This experience may provide some insight into cap-and-trade design issues as the United States debates the issue and some states and regions implement their own programs.

- The EU-ETS experience illustrates the importance of having reliable emissions data for all facilities covered under a cap-and-trade scheme; data that are pivotal for developing allocation systems, reduction targets, and enforcement provisions.
- In the United States debate continues on comprehensive versus sector-specific reduction programs; the EU-ETS experience suggests that adding sectors to a trading scheme once established may be a slow, contentious process.
- As with most EU industries, most US industry groups either oppose auctions outright or want them to be supplemental to a base free allocation. The EU-ETS experience suggests Congress and/or states may want to consider specifying any auction requirement if it wishes to incorporate market economics more fully into compliance decisions.

EU-ETS analysis suggests the most important variables in determining Phase 1 allowance price changes were oil and natural gas price changes; this apparent linkage raises possible market manipulation issues, particularly with the inclusion of financial instruments such as options and futures contracts. The EU will examine the matter in preparation for Phase 3. Congress may consider whether the government currently has sufficient regulatory and oversight authority over such instruments.

Chapter 2 - Beginning January 1, 2012, most carbon dioxide (CO_2) emissions from commercial flights to, from, and within the European Union (EU) are covered by the EU Emission Trading Scheme (EU ETS). Flights are covered regardless of whether the airline or operator is based in the EU region. The EU ETS caps aviation emissions of CO_2 in 2012 at 97% of the average in 2004-2006 and at 95% in each year from 2013 to 2020. Each April, beginning

2013, covered aircraft operators must turn in emission "allowances" (permits) equal to the previous year's emissions from their flights arriving at or departing from EU airports. Airline operators will receive free allowances for 82%- 85% of their 2010 emissions. Airlines that have more allowances than they need may sell them to others or save them for future use. Airlines that need more allowances may buy them from EU auctions, other carriers, other emission sources in the EU ETS, brokers, or international emission trading mechanisms. A small reserve of free allowances will be available for new or rapidly expanding airlines.

The entry into force of an EU law covering international aviation emissions is a significant move in a two-decade process concerning whether and how aviation emissions of CO_2 may be abated. Even among those who agree on the benefits of reducing greenhouse gas (GHG) emissions, how to share reductions across nations and sectors remains a continuing, major controversy internationally and within countries. Emissions from international air transport have, since early negotiations, posed a particular challenge.

The 1997 Kyoto Protocol (to which the United States is not a Party) specified that Parties should pursue limitation or reduction of GHG emissions from aviation fuels, working through the International Civil Aviation Organization (ICAO), an agency of the United Nations. In 2008, the EU cited a lack of "substantive progress" in ICAO and legislated to include aviation in its existing EU ETS by 2012. ICAO members have agreed to a variety of voluntary actions and goals. In October 2011, faced with impending inclusion of aviation emissions in the EU ETS, the ICAO Council agreed to accelerate its work, including continuing to explore market-based measures, CO_2 standards for new aircraft, and other options.

EU officials have stated their preference for agreeing on global, binding measures in ICAO. One official stated that the EU would agree to suspend inclusion of aviation in the ETS only if a new global ICAO scheme met three conditions: it must deliver more emissions reductions than the EU ETS on its own; it must have targets and measures; and any action must be non-discriminatory and apply to all airlines. Alternatively, EU law allows exemptions for incoming flights from countries that have adopted "equivalent measures" to reduce emissions. Some countries are likely negotiating with the EU for an equivalent measures exemption for their airlines.

Air carriers from the United States and other countries have vociferously objected to inclusion of international aviation in the EU ETS. The US government and other nations have pressed the EU to exclude foreign carriers. Two bills in Congress address the controversy. The US House of

Representatives passed H.R. 2594, the European Union Emissions Trading Scheme Prohibition Act of 2011. It would prohibit US aircraft operators from participating in the EU ETS. It also would direct the Administration to negotiate and take other actions to ensure that US civil aircraft operators are not penalized by any unilateral EU regulation of GHG emissions. A similar bill with more flexibility, S. 1956, was introduced in the Senate. In January 2012, House and Senate conferees on reauthorization of the Federal Aviation Administration (FAA) agreed to a sense of the Congress resolution opposing the EU action.

In: The European Union's Emissions … ISBN: 978-1-62081-955-5
Editors: A. R. Moore and M. F. Ryan

Chapter 1

CLIMATE CHANGE AND THE EU EMISSIONS TRADING SCHEME (ETS): LOOKING TO 2020*

Larry Parker

SUMMARY

The European Union's (EU) Emissions Trading Scheme (ETS) is a cornerstone of the EU's efforts to meet its obligation under the Kyoto Protocol. It covers more than 10,000 energy intensive facilities across the 27 EU Member countries; covered entities emit about 45% of the EU's carbon dioxide emissions. A "Phase 1" trading period began January 1, 2005. A second, Phase 2, trading period began in 2008, covering the period of the Kyoto Protocol. A Phase 3 will begin in 2013 designed to reduce emissions by 21% from 2005 levels.

Several positive results from the Phase 1 "learning by doing" exercise assisted the ETS in making the Phase 2 process run more smoothly, including: (1) greatly improving emissions data, (2) encouraging development of the Kyoto Protocol's project-based mechanisms—Clean Development Mechanism (CDM) and Joint Implementation (JI), and (3) influencing corporate behavior to begin

* This is an edited, reformatted and augmented version of a Congressional Research Service publication, CRS Report for Congress R41049, from www.crs.gov, prepared for Members and Committees of Congress, dated January 27, 2011.

pricing in the value of allowances in decision-making, particularly in the electric utility sector.

However, several issues that arose during the first phase were not resolved as the ETS implemented Phase 2, including allocation schemes and new entrant reserves, and others. A more comprehensive and coordinated response has been designed for Phase 3 with harmonized and coordinated rules being developed by the European Commission.

The United States is not a party to the Kyoto Protocol. However, six years of carbon emissions trading has given the EU valuable experience in designing and operating a greenhouse gas trading system. This experience may provide some insight into cap-and-trade design issues as the United States debates the issue and some states and regions implement their own programs.

- The EU-ETS experience illustrates the importance of having reliable emissions data for all facilities covered under a cap-and-trade scheme; data that are pivotal for developing allocation systems, reduction targets, and enforcement provisions.
- In the United States debate continues on comprehensive versus sector-specific reduction programs; the EU-ETS experience suggests that adding sectors to a trading scheme once established may be a slow, contentious process.
- As with most EU industries, most US industry groups either oppose auctions outright or want them to be supplemental to a base free allocation. The EU-ETS experience suggests Congress and/or states may want to consider specifying any auction requirement if it wishes to incorporate market economics more fully into compliance decisions.

EU-ETS analysis suggests the most important variables in determining Phase 1 allowance price changes were oil and natural gas price changes; this apparent linkage raises possible market manipulation issues, particularly with the inclusion of financial instruments such as options and futures contracts. The EU will examine the matter in preparation for Phase 3. Congress may consider whether the government currently has sufficient regulatory and oversight authority over such instruments.

OVERVIEW[1]

Climate change is generally viewed as a global issue, but proposed responses typically require action at the national level. With the 1997 Kyoto

Protocol now in force and setting emissions objectives for 2008-2012, countries that ratified the protocol are implementing strategies to begin reducing their emissions of greenhouse gases.[2] In particular, the European Union (EU) has decided to use an emissions trading scheme (called a "cap-and-trade" program), along with other market-oriented mechanisms permitted under the Protocol, to help it achieve compliance at least cost.[3] The decision to use emission trading to implement the Kyoto Protocol is at least partly based on the successful emissions trading program used by the United States to implement its sulfur dioxide (acid rain) control program contained in Title IV of the 1990 Clean Act Amendments.[4]

The EU's Emissions Trading System (ETS) covers more than 10,000 energy-intensive facilities across the 27 EU Member countries, including oil refineries, powerplants over 20 megawatts (MW) in capacity, coke ovens, and iron and steel plants, along with cement, glass, lime, brick, ceramics, and pulp and paper installations. In addition, aviation is currently being phased into the ETS. These covered entities emit about 40%-45% of the EU's total greenhouse gas emissions, and almost two-thirds of them are combustion installations. The trading program does not cover either carbon dioxide (CO_2) emissions from the transportation sector (except aviation), which account for about 25% of the EU's total greenhouse gas emissions, or emissions of non-CO_2 greenhouse gases, which account for about 20% of the EU's total greenhouse gas emissions. A Phase 1 trading period ran between January 1, 2005, and December 31, 2007.[5] A Phase 2 trading period began January 1, 2008, covering the period of the Kyoto Protocol, and a Phase 3 has been finalized to begin in 2013.[6]

Under the Kyoto Protocol, the then-existing 15 nations of the EU agreed to reduce their aggregate annual average emissions for 2008-2012 by 8% from the Protocol's baseline level (mostly 1990 levels) under a collective arrangement called a "bubble." In light of the Kyoto Protocol targets, the EU adopted a directive establishing the EU-ETS that entered into force October 13, 2003.[7]

One objective of the second phase of the ETS is to achieve 3.3 percentage points of the 8.0% reduction required by the EU-15 under the Protocol.[8]

The importance of emissions trading was elevated by the accession of 12 additional central and eastern European countries to EU membership from May 2004 through January 2007. For the new EU-27, the overall ETS emissions cap is set at 2.08 billion metric tons of carbon dioxide (CO2) annually for the Kyoto compliance period (2008-2012).

The second phase Kyoto compliance stage of the ETS is built on the experience the EU gained from its preliminary Phase 1. The European Commission (EC) believes that the Phase 1 "learning by doing" exercise prepared the community for the difficult task of achieving the reduction requirements of the Kyoto Protocol. Several positive results from the Phase 1 experience assisted the ETS in making the Phase 2 process run smoothly, at least so far. First, Phase 1 established much of the critical infrastructure necessary for a functional emission market, including emissions monitoring, registries, and inventories. Much of the publicized difficulty the ETS experienced early in the first phase can be traced to inadequate emissions data infrastructure.[9]

Phase 1 significantly improved those critical elements in preparation for Phase 2 implementation.

Second, the ETS helped jump-start the project-based mechanisms—Clean Development Mechanism (CDM) and Joint Implementation (JI)—created under the Kyoto Protocol.[10] As stated by Ellerman and Buchner:

> The access to external credits provided by the Linking Directive has had an invigorating effect on the CDM and more generally on CO2 reduction projects in developing countries, especially in China and India, the two major countries that will eventually have to become part of a global climate regime if there is to be one.[11]

Third, according to the EC, a key result of Phase 1 was its effect on corporate behavior.

An EC survey of stakeholders indicated that many participants are incorporating the value of allowances in making decisions, particularly in the electric utility sector, where 70% of firms stated they were pricing the value of allowances into their daily operations, and 87% into future marginal pricing decisions. All industries stated that it was a factor in long-term decision-making.[12]

However, several issues that arose during the first phase remained contentious as the ETS implemented Phase 2, including allocation (including use of auctions and reliance on model projections), new entrant reserves, and others.

In addition, the expansion of the EU and the implementation of the linking directives created new issues to which Phase 2 has had to respond.[13] Based on lessons learned in Phase 1 and Phase 2, the EU has taken a substantially different approach to these issues in Phase 3 that is discussed later.

RESULTS FROM PHASE 1 AND 2

It is unclear to what degree the first phase of the ETS achieved real emissions reductions. Emissions are dynamic over time; a product of a country's population, economic activity, and greenhouse gas intensity.[14] To capture these dynamics, each Member State of the EU developed an emissions baseline from models that project future trends in the country's emissions based on these and other factors, such as anticipated energy and greenhouse gas policies.[15] During the first phase, the emissions goal was to put the EU on the path to Kyoto compliance—not actually comply with the Protocol (which wasn't necessary until the 2008-2012 time period). Thus, countries developed "business as usual" baselines based on projected growth in emissions. Such a projected baseline suffers from two sources of uncertainty: data uncertainties, and forecasting uncertainties. On data, Phase 1 suffered from uncertainties with respect to data collection and coverage, in monitoring methods for historic data, and data verification. On projecting future emissions, Phase 1 faced uncertainties with respect to economic or sector-based growth rates. Fueled in many cases by over-optimistic economic growth assumptions, these uncertainties increased the probability of inflated business as usual baselines.[16]

The combination of these factors and modest reduction requirements resulted in the emissions allocations for the 2005-2007 trading period being higher than actual 2005 emissions.[17] This result raised questions about how much reductions achieved during Phase 1 were real as opposed to being merely paper artifacts. On the positive side, verified emissions in 2005 were 3.4% below the estimated 2005 baseline used during the allocation process. In addition, the allowance prices for 2005 stayed persistently high, suggesting some abatement was occurring and raising questions of "windfall" profits. As stated by Ellerman and Buchner:

> First, and most importantly, the persistently high price for EUAs [EU emissions allowances] in a market characterized by sufficient liquidity and sophisticated players must be considered as creating a presumption of abatement. It would be startling if power companies did not incorporate EUA prices into dispatch decisions that would have shifted generation to less emitting plants. There is plenty of anecdotal evidence that this was the case, and the prominent charges of windfall profits assume that the opportunity cost of freely allocated allowances was being passed on (without noting the implications for abatement). Similarly, it would be

surprising if there were no changes in production processes that could be made by the operators of industrial plants.[18]

However, EU emissions allowances (EUAs) during Phase 1 did not maintain value. Phase 1 EUAs were basically worthless during the final six months of 2007. This decline in EUA prices at least partially reflected the general non-transferability of Phase 1 EUAs to Phase 2. Only Poland and France included limited banking in their Phase 1 implementation plans (called National Allocation Plans (NAPs)). The EC further restricted use of Phase 1 EUAs in Phase 2 with a ruling in November 2006.[19] As a result, excess Phase 1 EUAs were worthless at the end of 2007.[20]

One consequence of the non-transferability of Phase 1 EUAs is that prices for Phase 2 EUAs remained relatively firm until the 2008-2009 recession reduced demand, as indicated by *Figure 1*.[21] Scarcity is critical for the proper functioning of an allowance market. As further indicated by *Figure 1*, during 2009, the market firmed up at a much lower level as participants assessed the impact of the recession on the demand for EUAs. Looking at 2010 data for Phase 2 allowances (December 2010 settlement date) indicate prices staying in the range of 13-16 euro as the European economy slowly recovers from the recession. This is a different response than the market had during Phase 1, and may reflect Phase 2 improvements in the system. In particular, the more predictable 2009-2010 response may reflect the ability of the EC to certify Phase 2 NAPs using more verifiable baseline data than were available for Phase 1.[22] A major reason the EC rejected *ex post* adjustments[23] was fear that such adjustments would have a disruptive effect on the marketplace.[24] Phase 1 did not firmly establish this foundation of markets;[25] based on the Phase 2 EUA future's market, further market development appears to be occurring, although, like most commodity markets, it remains somewhat volatile at times.

While the environmental performance of Phase 1 may be disputed, the need for additional reductions to achieve Kyoto is not. Projections conducted by the European Environment Agency (EEA) in 2009 indicated that for 2008, the EU-15 was 6.2% below its base-year emissions, compared with an 8% five-year average reduction commitment under the Kyoto Protocol.[26] As indicated by *Figure 2*, this represents a continuation of reductions by EU-15 over the past five years. However, as indicated by the pink line, the EEA projects that the EU-15 existing measures are insufficient to reduce EU-15 emissions to their Kyoto requirements (represented by the purple line), resulting in a projected 6.9% reduction from baseline levels. To achieve the Kyoto target the EU projects further reduction actions planned by EU-15 countries (represented by the green line in *Figure 2*), resulting in an overall reduction of 8.5% compared with baseline levels.

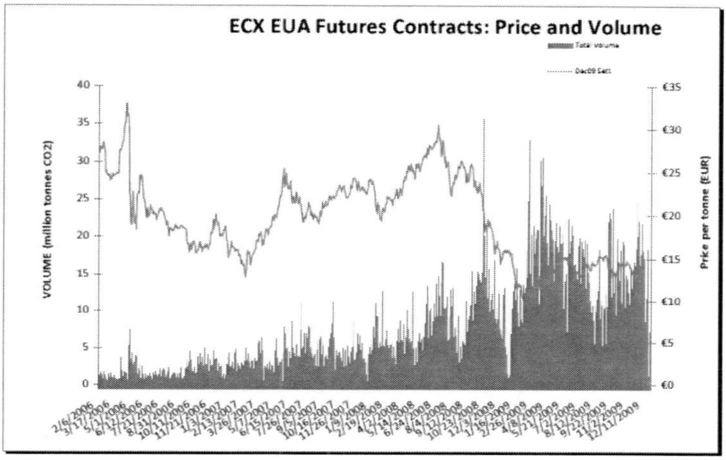

Source: ECX Exchange.

Note: Dec09 Sett: Future contracts with a settlement date of December, 2009.

Figure 1. ECX CFI Futures Contracts: Price and Volume.

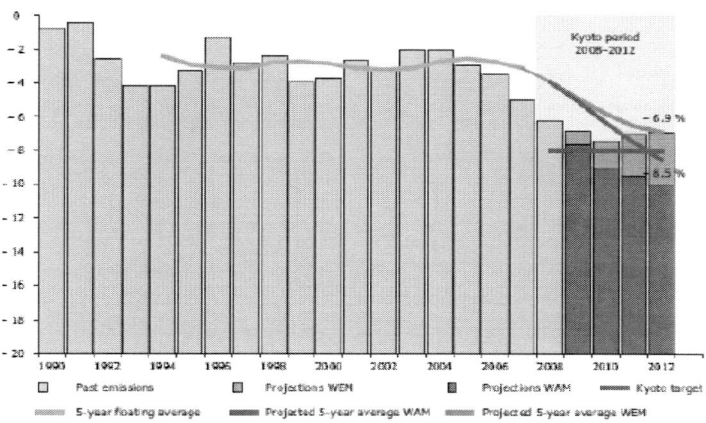

Source: European Environment Agency, Greenhouse Gas Emissions Trends and Projects in Europe 2009, (2009) p.10.

Note: WEM: with existing measures (measures implemented or adopted). WAM: with additional measures (planned measures).

Figure 2. EU-15 Greenhouse Gas Emissions and Projections for the Kyoto Period: 1990-2012.

In addition to domestic emission reductions, the EU has also projected additional reduction credits received by activities permitted under the Kyoto Protocol: (1) purchase of project-based credits by ETS participants and EU governments (e.g., Joint Implementation (JI) and Clean Development Mechanism (CDM) projects); and, (2) the use of carbon sinks. As indicated in *Figure 3*, these activities provide a credit on the EU-15 baseline of 4.6 percentage points.

Thus, if the EU-15 maintains its current path, it would exceed its Kyoto commitment by about 3.5 percentage points (6.9% minus 3.4%). If its planned measures result in the projected 8.5% reduction below baseline levels, the overachievement of its Kyoto commitment would be 5.1 percentage points (8.5% minus 3.4%).[27]

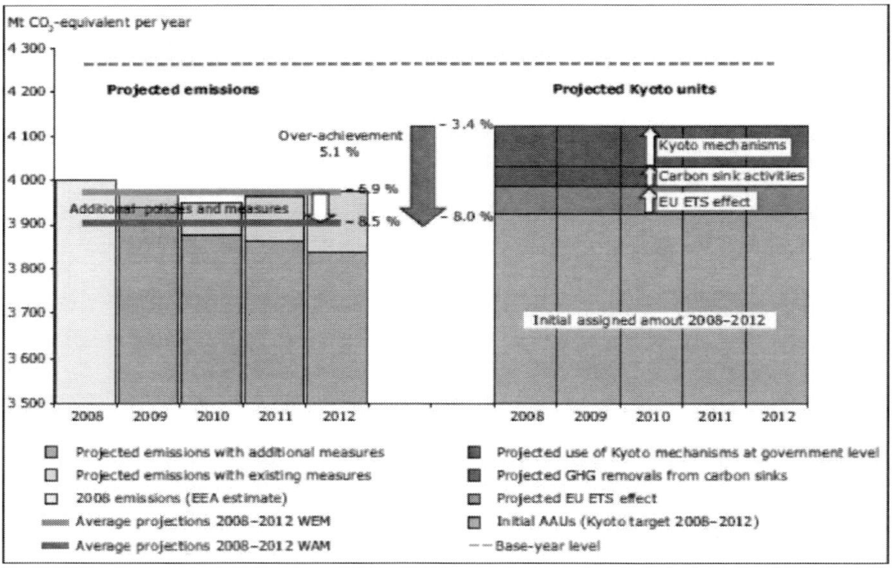

Source: European Environment Agency, Greenhouse Gas Emissions Trends and Projects in Europe 2009, (2009) p. 11.

Notes: The left section shows the projected emissions considering only domestic measures (existing and additional) and is showing them as average 2008-2012 emissions (lines) and annual emissions (bars). The right section shows the projected amount of Kyoto credits accumulated by the end of the commitment period, including the initial EC assigned amount under the Protocol, the purchase of Kyoto project credits by EU ETS participants and EU governments, and carbon sink activities.

Figure 3. Summary of EU-15 Emissions Projection Compared to Projected Kyoto Protocol Credits.

The EU-27 as a whole does not have an emissions target comparable to the EU-15 bubble. By 2010, EU-27 emissions are projected at 9.6% below Kyoto baseline levels assuming current policies. This reduction is projected at 11.3% if additional measures are included.

Based on EEA's 2009 projections, 24 of the 25 countries with reduction requirements are projected to meet their commitments under the Kyoto Protocol.[28] Only Austria is not projected to meet its requirements even with additional planned measures and the use of Kyoto mechanisms.[29] After the first year of the Kyoto compliance period, seven EU-15 member countries are already in compliance with their 2008-2012 target: Belgium, Finland, France, Germany, Greece, Sweden, and the United Kingdom. For the ten of EU-12 member countries with Kyoto obligations, only Slovenia was not in compliance in 2008.[30] Total EU-27 greenhouse gas emissions in 2008 were 4.94 billion metric tons, compared with 5.57 billion metric tons in 1990 – a drop of 11.3%.

PHASE 3

The European Union is committed to achieving a 20% reduction in greenhouse gas emissions by 2020 from 1990 levels (or more depending on the actions of other countries). A strategic component of the effort to achieve this target is a revised ETS that will achieve a 21% reduction from covered entities from 2005 levels. *Table 1* indicates the proposed EU-wide ETS cap for the next Phase of EU greenhouse gas program (Phase 3) assuming no further international commitments. As indicated, the EC envisions a linear reduction in the ETS cap to match the reduction target under the overall 20% reduction program. These numbers will change as individual countries decide to include more facilities under the ETS and as the EC expands ETS coverage to include other sectors and non-CO_2 greenhouse gases.

In July, 2010, the EC issued a decision on the EU emissions allowance cap under the EU-ETS for 2013. Set at 2,032,998,912 allowances, the cap is subject to further revisions if additional relevant information warrants a change and when a separate decision on allocations to aircraft operators is made.[31]

For Phase 3, the EU is re-shaping the ETS to improve its efficiency and eliminate some of the problems identified during Phase 1 and 2.[32] For Phase 2, the improved emissions inventories resulting from Phase 1 allowed the EC to harmonize the types of installations covered by the ETS across the various Member States.[33] In addition, the EC imposed a uniform rule on the Member

States preventing the use of *ex-post* adjustments. However, Phase 2 made little advancement in harmonizing individual countries' allocations schemes.[34] As with Phase 1, countries continue to differ widely on several key points.

Table 1. Proposed Annual ETS Cap Figures for Phase 3

Year	Billion metric tons of CO_{2e}
Annual limit for Kyoto compliance period	2.083
(2008-2012)	
2013	1.974
2014	1.937
2015	1.901
2016	1.865
2017	1.829
2018	1.792
2019	1.756
2020	1.720

Source: European Commission, Questions and Answers on the Commission's Proposal to revise the EU Emissions Trading System (Brussels, January 23, 2009), response to question 12.

Note: Figures are based on the current Phase 2 scope of the ETS. These need to be adjusted for three reasons: (1) extensions of ETS scope during phase 2 by Member states; (2) extensions of ETS scope by the EC for third trading period, and (3) the figures do not include inclusion of aviation, nor the emissions from Norway, Iceland, and Liechtenstein—non-EU countries that have linked their programs to the ETS.

The critical structural change the EU would make in Phase 3 is eliminating National Allocation Plans (NAPs), and replacing them with EU-wide rules with respect to allowance availability, allocations, and auctions. NAPs are central to the EU's effort to achieve its Kyoto obligations under Phase 2. Each Member of the EU submitted a NAP that lays out its allocation scheme under the ETS, including individual allocations to each affected unit. These NAPs were assessed by the EC to determine compliance with 12 criteria delineated in an annex to the emissions trading directive.[35] Criteria included requirements that the emissions caps and other measures proposed by the Member State were sufficient to put it on the path toward its Kyoto target, protections against discrimination between companies and sectors, and delineation of intended use of CDM and JI credits for compliance, along with provisions for new entrants, clean technology, and early reduction credits. For the second trading period, the NAP had to guarantee Kyoto compliance.

This NAP structure will be replaced under Phase 3. There will be one EU-wide cap instead of the 27 national caps under Phase 1 and 2. Allowances will be allocated under EU-wide, fully harmonized rules, including those governing: (1) auctions, (2) transitional free allocations for greenhouse gas intensive, trade-exposed industries, (3) new entrants, and (4) coverage. The EC proposed a Directive to alter the EU-ETS structure for Phase 3 in January, 2008, [36] and the Directive was amended and adopted by the European Parliament (EP) and of the Council of the European Union in April 2009.[37]

Auctions

Under Phases 1 and 2, allowances generally were and are allocated free to participating entities under the ETS. During Phase 1, The 2003 EU-ETS Directive allowed countries to auction up to 5% of allowance allocations, rising to 10% under Phase 2.[38] Under Phase 1, only 4 of 25 countries used auctions at all, and only Denmark auctioned the full 5%. The political difficulty in instituting significant auctioning into ETS allowance allocations is the almost universal agreement by covered entities in favor of free allocation of allowances and opposition to auctions.[39] Free allocation of allowances represents a one-time transfer of wealth to the entities receiving them from the government issuing them.[40] The resulting transfer of wealth has been described by several analysts as "windfall profits."[41] As summarized by Ellerman and Buchner: "Allocation in the EU ETS provides one more example that, notwithstanding the advice of economists, the free allocation of allowances is not to be easily set aside."[42]

Despite concerns about windfall profits and economic distortions resulting from the free allocation of allowances, there was little change in basic allocation philosophy for Phase 2. No country proposed auctioning the maximum percentage of allowances allowed (10%). Most do not include auctions at all.[43] The unwillingness of governments to employ auctions as an allocating mechanism revolve around equity considerations, including: (1) inability of some covered entities to pass through cost because of regulation or exposure to international competition; (2) potential drag on a sector's economic performance from the up-front cost of auctioned allowances; and (3) the potential that government will not recycle revenues to alleviate compliance costs, international competitiveness impacts, or other equity concerns, resulting in the auction costs being the same as a tax.[44]

This opposition is mostly overcome for Phase 3 through an EU-wide set of harmonized rules for allowance allocations and auctions. Under Phase 3, the Directive states:

> Auctioning should ... be the basic principle for allocation, as it is the simplest, and generally considered to be the most economic efficient system. This should also eliminate windfall profits and put new entrants and economies growing faster than average on the same competitive footing as existing installations. (paragraph 15)

After nine eastern European Member States threatened to veto an initial proposal to auction 100% of all allowances, the EU compromised to provide for some free allocation of allowances during Phase 3 that will begin in 2013.[45]

Most covered industries will be eligible for some free allocation of allowances to cover direct emissions under the Phase 3 agreement. The introduction of auction would be differentiated by sector. In general, for the power sector, full auctioning will begin in 2013. For electric powerplants, most will receive no free allocation of allowances during Phase 3. However, in a concession to certain eastern European Member States, an optional and temporary derogation from the no-free-allocation requirement for powerplants is provided to countries that meet specific energy and economic criteria. Under the optional allocation scheme, the Member State can allocate allowances equal to 70% of the powerplant's Phase 1 emissions free; this allocation declines to zero in 2020.

The auction schedule for most other covered entities is more gradual with 80% of a sector's allocation provided free in 2013, declining linearly to 30% by 2020, and zero by 2027. As stated in the Directive:

> For other sectors covered by the Community scheme, a transitional system should be foreseen for which free allocation in 2013 would be 80% of the amount that corresponded to the percentage of the overall Community-wide emissions throughout the period 2005 to 2007 that those installations emitted as a proportion of the annual Community-wide total quantity of allowances. Thereafter, the free allocation should decrease each year by equal amounts resulting in 30% free allocation in 2020, with a view to reaching no free allocation in 2027. (paragraph 21)

Because of concern that stringent EU carbon policies may encourage production and related greenhouse gas emissions to shift to countries without carbon policies (i.e., carbon leakage), exceptions to this phase-out of free allowances will be made in sectors where carbon leakage may occur, as discussed later. Distribution of allowances to be auctioned by the Member States will be determined by a three-part formula (Article 10(2)). Eighty-eight percent of the allowances to be auctioned by each Member State is distributed to States according to their historic emissions under Phase 1 of the EU-ETS. Ten percent of the total is distributed to States mostly based in their comparative GDP per capita within the EU (Annex IIa). Two percent of the total is distributed to nine former eastern-bloc countries based on the substantial greenhouse gas reductions they have already achieved (Annex IIb).

In November 2010, the EC finalized an Auction Regulation to provide for coordinated auctions of allowances during Phase 3.[46] The rule creates a central platform to auction off the majority of allowances, beginning in 2013. While the EC would have preferred to have a single platform, some members insisted on having the option of opting out of the central platform and setting up their own platforms under specified conditions. Allowances will be auctioned on a spot basis, avoiding potential conflicts with established secondary markets (such as the ECX). The format will be a single-round, sealed bid, uniform price auction. Auctions will be held at least weekly.

Auctions will be open to facility operators, and regulated investment firms and credit institutions, along with "intermediaries (such as fuel traders). The rule also includes provisions to address potential anti-competitive behavior and market abuse.

Beyond the allocation of allowances, the EU Directive also provides guidelines for the allocation of revenues from allowance auctions. The Directive states that at least 50% of the proceeds should be used to fund a variety of climate change related activities, including emission reductions, adaptation activities, renewable energy, carbon capture and storage (CCS), the Global Energy Efficiency and Renewable Energy Fund, and assisting developing countries to avoid deforestation and increase afforestation and reforestation (Article 10(3)).

New Entrant Reserves

Unlike previous cap-and-trade programs, the EU-ETS includes provisions for allocating free allowances to new entrants to the system.[47] The reasoning behind this decision is based on equity: (1) it isn't fair to allocate allowances free to existing entities while requiring new entrants to purchase them, and (2) the EU doesn't want to put Member States at a disadvantage in competing for new investments.[48] These equity concerns trumped concerns about economic efficiency.

As is the case for existing entities, the free allocation of allowances to new entrants is a subsidy. Under Phase 1 and Phase 2, the size and distribution of this subsidy is left to the individual Member States. For Phase 1, the reserve varied widely from the average of 3% of total allowances: Poland set aside only 0.4% of its allocation for new entrants while Malta set aside 26%. For Phase 2, the spread continues with Poland reserving 3.2% of its allowances for new entrants in contrast to 45% reserved by Latvia.[49]

The decision to employ a new entrant reserve adds complexity to Member States' allocation plans and influences the investment decisions of covered entities. Rules had to be promulgated with respect to the reserve's size, manner in which the allowances are dispensed, and how to proceed if the demand either exceeds the supply, or vice versa. As indicated, countries did not harmonize new entrant reserve rules with respect to size during Phase 1 or 2. Likewise, there is no standardization on dispensing allowances and replenishing the reserve: first-come, first-serve with no replenishment is one approach used, but a variety of procedures have been developed both to dispense allowances and to replenish the reserve if supply is inadequate. Member States also have different formulas for determining how many allowances a new entrant should receive. Member States claim to use a form of "benchmarking" to determine allowance allocations—an approach based on a standard of "best practices" or "best technology" that is applied to the new entrant's anticipated production or capacity. However, the definitions and application of the benchmarks used by the Member States are not uniform.

This will change under Phase 3. Under Phase 3, the Directive sets an EU-wide cap of 5% of the total allowance cap for a new entrant reserve, and requires the harmonization of allocation rules. The EC is to adopt a harmonized rule for applying a new entrant definition contained in the Directive by December 31, 2010; the Directive expressly excludes any new electricity production from being defined as a new entrant. The EC is also to determine EU-wide benchmarks for the allocation of all free allowances. The Directive states that the starting point for setting those benchmarks shall be the average performance of the 10% most efficient installations in a sector or subsector in the EU in the years 2007-2008 (Article 10a(2)).

In an attempt to stimulate development of CCS, the Directive also provides that up to 300 million allowances in the new entrants' reserve shall be available through 2015 for aiding construction and operation of up to 12 demonstration projects.

No one project can receive more than 15% of the allowances allocated for this purpose (Article 10a(8)).

Draft rules to harmonized free allocation of allocations through benchmarks was proposed by the EC on December 15th, 2010.[50] As drafted, a benchmark represents the threshold for receiving free allocations for a specific facility; it does not represent an emission limit or even an emission reduction target.

The benchmarks are to be developed per product, to the extent feasible. As noted in the Directive, a product benchmark generally is based on the average greenhouse gas performance of the 10 % best performing installations in the EU producing that product.

As explained by the EC: "The benchmarks are established on the basis of the principle 'one product = one benchmark', which means that the benchmark methodology does not differentiate by technology or fuel used, nor the size of an installation or its geographical location."

EC Phase 3 Decision on Eligible Industries[51]

Most studies of the competitiveness impacts of the ETS during Phase 1 have found no impact. The International Energy Agency (IEA) cites several reasons for this situation:

> Experience to date with the EU-ETS does not reveal leakage for the sectors concerned— analysis of steel, cement, aluminum and refineries sectors reveals that no significant changes in trade flows and production patterns were evident during the first phase (2005-2007) ofthe EU-ETS. This is mostly due to the free allocation of allowances, sometimes in generous quantities, and to the still functioning long-term electricity contracts, which softened the blow of rising electricity prices. Further, the general boom in prices for most traded products subject to carbon costs—whether direct or indirect—has blurred any effects of the latter. Finally, the relatively short time span of these policies does not allow observation of the full potential effects on industry via changes in investment location decisions.[52]

This conclusion is echoed by Carbon Trust, which states that currently, free allocation of emissions allowances offset almost all of the additional costs of the ETS; and that conclusion is echoed by The Climate Group for The German Marshall Fund, which states that companies surveyed found it difficult to quantify effects on their bottom line in the first phase, or found no effect at all.[53]

For energy-intensive, trade-exposed industries, Phase 3 has provisions to provide assistance to eligible installations to address the direct and indirect impact of emissions control costs.

With respect to direct emissions costs, the EC published a list of installations exposed to a significant risk of carbon leakage on December 24, 2009, as required under the Directive.[54] The list is identical to the draft list released in September 2009.[55]

The decision lists 164 industrial sectors and subsectors deemed to be exposed sectors under the appropriate European Parliament and Council directives.

Eligible installations will receive allowances sufficient to cover 100% of their direct emissions, provided they are using the most efficient technology available (based on the benchmarks provided in the draft rules discussed above).

This 100% allocation contrasts with the 80% distribution of free allowances to non-carbon leakage exposed industries in 2013. Reflecting the fluid nature of the competitive situation and international negotiations, the EC was to review its decision June 30, 2010, and provide the European Parliament and Council with any appropriate proposals to respond to the situation. Consultations made by the EC with interested parties suggest the international situation has not changed sufficiently for the EC to propose changes to its Directive.

Assistance for the impact of indirect emissions control costs on exposed industries from higher electricity prices would be determined by Member States. As stated by the Directive:

> Member States may deem it necessary to compensate temporarily certain installations which have been determined to be exposed to a significant risk of carbon leakage related to greenhouse gas emissions passed on in electricity prices for these costs. Such support should only be granted where it is necessary and proportionate and should ensure that the Community scheme incentives to save energy and to stimulate a shift in demand from grey to green electricity are maintained. (paragraph 27)

Flexibility Mechanisms and Price Volatility Control

The major flexibility mechanism developed under the EU-ETS has been the Clean Development Mechanism (CDM) and Joint Implementation (JI) credits permitted under the Kyoto Protocol; however, this development has proven a controversial process. A major part of the controversy has been the "supplementarity" requirement of the Kyoto Protocol to use its flexibility mechanisms. Supplementarity requires that developed countries, such as most EU countries, ensure that their use of JI/CDM credits is supplemental to their own domestic control efforts. In defining supplementarity for Phase 2, the EC used 10% of a country's allowance allocation as a rule of thumb in approving NAPs—with a greater limit possible based on a country's domestic efforts to reduce emissions. This process resulted in some significant reductions in some countries' proposed limits (e.g., Ireland, Poland, Spain), but some increase in others (e.g., Italy, Latvia, Lithuania). Although these reductions appear substantial in individual cases, most analysts agree that they do not represent a major barrier to the cost-effective use of JI/CDM. However, the EU-ETS does not accept credits from land use, land-use change and forestry (LULUCF) projects.

For Phase 3, the EU maintains its ban on using LULUCF credits within the ETS. In addition, in November 2010, the EC proposed rules to ban the use of credits from two industrial gases—a rule likely to be adopted formally by the EC in Spring 2011. The ban will apply to projects which destroy trifluoromethane (HFC-23) produced as a by-product of chlorodifluoromethane (HCFC22) production, and nitrous oxide (N2O) from adipic acid production. According to the EC, just 23 such industrial gas projects account for two-thirds of all the credits generated through the CDM; most are in China and other advanced developing countries. The ban on the use of such credits in the EU ETS will apply to all such projects undertaken under the CDM as well as any undertaken in developed countries through the JI mechanism. Essentially, the ban allows companies to use for 2012 compliance under the EU ETS, but not thereafter.[56]

However, Phase 3 will permit up to 50% of the required reductions mandated under Phase 3 to be achieved through CDM or JI credits. For existing installations, this represents a total of 1.6 billion credits over the eight-year compliance period. Limits on use of Kyoto credits will be based on a facility's 2008-2012 allocation (for an existing facilities) or its verified emissions during Phase 3 (for a new entrant or sector). The EC estimates that the minimum amount of Kyoto credits an existing facility will be able to use to comply with Phase 3 will be 11% of its 2008-2012 allocation, while new entrants and sectors will be able to use a minimum of 4.5% of their verified emissions during 2013-2020 (article 11a(8)). The precise percentages will be determined later.

Another flexibility mechanism, banking, is extended by the Directive from Phase 2 to Phase 3 in order to prevent a Phase 1 style collapse of allowance prices when the ETS transitions into Phase 3. In addition, the EU hopes that extending the trading period from five years to eight years, along with the steady, linear emissions reduction schedule, will increase certainty and stability in the allowance markets.

Phase 3 will introduce two other mechanisms designed to address price volatility. First, the EC is required under the Directive to examine whether the market for emission allowances is sufficiently protected from insider dealing or market manipulation. If not, the EC is to present proposals to ensure such protection to the EP and the Council (article 12(a)).

Second, the Directive provides that if the allowance price is more than three times the preceding two-year average for more than six consecutive months and the price is not based on market fundamentals, one of two measures may be taken. The first would allow Member States to shift forward the auctioning of some of its auctionable allowances. The second would allow Member States to auction up to 25% of the remaining allowances in the new entrants reserve (article 29(a)).

Expanding Coverage

Despite the EC's interest in expanding the ETS, its coverage in terms of industries included for Phase 2 is essentially the same as for Phase 1. The exception is for aviation. In December, 2006, the EC proposed bringing greenhouse gas emissions from civil aviation into the ETS in two phases.[57] As agreed to by the European Parliament in July 2008, all intra-EU and international flights will be included under the ETS beginning in 2012. Emissions would be capped at 97% of average 2004-2006 emissions with 85% of the allowances being allocated free to operators. The cap would be reduced to 95% in 2013. The cap and auctioning of allowances would be reviewed as a part of Phase 3 implementation.

Annex I of the Directive identifies three CO2 emitting sectors for inclusion under the ETS: petrochemicals, ammonia, and aluminum. The ETS will also expand beyond CO2 to include nitrous oxide (N2O) emissions from nitric, adipic, and glyoxalic acid production, and perofluorocarbon (PFC) emissions from the aluminum sector. This would expand ETS covered emissions by 4.6% over Phase 2 allowance allocations, or about 100 million metric tons.[58] The harmonization and codification of eligibility criteria for combustion installations is expected to increase the coverage by a further 40-50 million metric tons.

To improve the cost-effectiveness of the ETS and reduce administrative costs, the Directive provides that small installations may be subject to other control regimes (such as carbon taxes) rather than included under the EU-ETS. Currently, the smallest 1,400 (10% of total installations covered) installations emit only 0.14% of total emissions covered. The Directive provides that Member States may opt to exclude installations that emit less than 25,000 metric tons annually from the EU-ETS (paragraph 11).

SUMMARY AND CONSIDERATIONS FOR US CAP-AND-TRADE PROPOSALS

The United States is not a party to the Kyoto Protocol and no legislative proposal before the Congress would impose as stringent or rapid an emission reduction regime on the United States as Kyoto would have. Likewise, US proposals to reduce emissions through 2020 are not as stringent as that provided in the EU Directive. However, through five years of carbon emissions trading, the EU has gained valuable experience. This experience, along with the process of developing Phase 3, may provide some insight into current cap-and-trade design issues in the United States.

Emission Inventories and Target Setting

The ETS experience with market trading and target setting confirms once again the central importance of a credible emissions inventory to a functioning cap-and-trade program.[59] The lack of credible EU-wide data on emissions was a direct cause of the ETS Phase 1 allowance market collapse in 2006. Arguably, the most important result of Phase 1 was the development of a credible inventory on which to base future targets and allocations.

In the United States, Section 821 of the 1990 Clean Air Act Amendments requires electric generating facilities affected by the acid rain provisions of

Title IV to monitor carbon dioxide in accordance with EPA regulations.[60] This provision was enacted for the stated purpose of establishing a national carbon dioxide monitoring system.[61] As promulgated by EPA, regulations permit owners and operators of affected facilities to monitor their carbon dioxide emissions through either continuous emission monitoring (CEM) or fuel analysis.[62] The CEM regulations for carbon dioxide are similar to those for the acid rain program's sulfur dioxide CEM regulations. Those choosing fuel analysis must calculate mass emissions on a daily, quarterly, and annual basis, based on amounts and types of fuel used.

As suggested by the EU-ETS experience, expanding equivalent data requirements to all facilities covered under a cap-and-trade program would be the foundation for developing allocation systems, reduction targets, and enforcement provisions. In response to FY2008 Consolidated Appropriations Act,[63] EPA promulgated the Greenhouse Gas Reporting Rule in 2009 to expand the facilities required to report greenhouse gas emission data to EPA annually.[64] Under the rule, 31 categories of sources begin reporting their emissions in 2011 if a source within a designated category emits 25,000 metric tons or more of carbon dioxide equivalents annually. Since promulgation, four additional categories have been added and others proposed.

Coverage

Despite economic analysis to the contrary, the EU decided to restrict Phase 1 ETS coverage to six sectors that represented about 40%-45% of the EU's CO_2 emissions.[65] This restriction was estimated to raise the cost of complying with Kyoto from 6 billion euro annually to 6.9 billion euro (1999 euro) compared with a comprehensive trading program. A variety of practical, political, and scientific reasons were given by the EC for the decision.[66]

The experience of the ETS up to now suggests that adding new sectors to an existing trading program is a difficult process. As noted above, a stated goal of the EC is to expand the coverage of the ETS. However, the experience of Phase 1 did not result in the addition of any new sector until the last year of Phase 2 when aviation will be included. The EU will expand its coverage with Phase 3, but the ETS will still cover fewer sectors emitting greenhouse gases than provided under most US proposals.

US cap-and-trade proposals generally fall into one of two categories. Most bills are more comprehensive than the ETS, covering 80% to 100% of the country's greenhouse gas emissions. At a minimum, they include the electric utility, transportation, and industrial sectors; disagreement among the bills center on the agricultural sector and smaller commercial and residential sources. In some cases discretion is provided EPA to exempt sources if serious data, economic, or other considerations dictate such a resolution.

A second category of bills focuses on the electric utility industry, representing about 33% of US greenhouse gases and therefore less comprehensive than the ETS. Sometimes including additional controls on non-greenhouse gas pollutants, such as mercury, these bills focus on the sources with the most experience with emission trading and the best emissions data. Other sources could be added as circumstances dictate.

As noted, the EU's experience with the ETS suggests that adding sectors to an emission trading scheme can be a slow and contentious process. If one believes that the electric utility sector is a cost-effective place to start addressing greenhouse gas emissions and that there is sufficient time to do the necessary groundwork to eventually add other sectors, then a phased-in approach may be reasonable. If one believes that the economy as a whole needs to begin adjusting to a carbon-constrained environment to meet long term goals, then a more comprehensive approach may be justified. The ETS experience suggests the process doesn't necessarily get any easier if you wait.

Allocation Schemes

Setting up a tradeable allowance system is a lot like setting up a new currency.[67] Allocating allowances is essentially allocating money with the marketplace determining the exchange rate. As noted above, the free allocation scheme used in the ETS has resulted in "windfall profits" being received by allowance recipients. As stated quite forcefully by Deutsche Bank Research:

> The most striking market outcome of emissions trading to date has been the power industry's windfall profits, which have sparked controversy. We are all familiar with the background: emissions allowances were handed out free of charge to those plant operators participating in the emissions trading scheme. Nevertheless, in particular the producers of electricity succeeded in marking up the market price of electricity to include the opportunity-cost value of the allowances. This is correct from an accounting point of view, since the allowances do have a value and could otherwise be sold. Moreover, emissions trading cannot work without price signals.[68]

The free allocation of allowances in Phase 1 and 2 of the ETS incorporates two other mechanisms that create perverse incentives and significant distortions in the emissions markets: new entrant reserves and closure policy. Combined with an uncoordinated and spotty benchmarking approach for both new and existing sources, the result is a greenhouse gas reduction scheme that is influenced as much or more by national policy than by the emissions marketplace. The expansion of auctions for Phase 3 of the ETS could simplify allocations and permit market forces to influence compliance strategies more fully. Most countries did not employ auctions at all during Phase 1 and auctions continue to be limited under Phase 2. No country combined an auction with a reserve price to encourage development of new technology. The EC limited the amount of auctioned allowances to 10% in Phase 2: a limit no country chose to meet. Efforts to expand auctions met opposition from industry groups, but attracted support from environmental groups and economists. The Phase 3 increased use of auctioning through 2020 will represent a major development for the scheme.

Currently, all US cap-and-trade proposals have some provisions for auctions, although the amount involved is sometimes left to EPA discretion. Most specify a schedule that provides increasing use of auctions from 2012 through the mid-2030s with a final target of 66%-100% of total allowances auctioned. Funds would be used for a variety of purposes, including programs to encourage new technologies. Some proposals include a reserve price on some auctions to create a price floor for new technology.

Like the situation in the ETS, most US industry groups either oppose auctions outright or want them to be supplemental to a base free allocation. Given the experience with the ETS where the EC and individual governments have been unwilling or unable to move away from free allocation, the Congress, like the EU, may ultimately be asked to consider specifying any auction requirement if it wishes to incorporate market economics more fully into compliance decisions.

Flexibility and Price Volatility

Despite EU rhetoric during the Kyoto Protocol negotiations, it moved into Phase 2 without a significant restriction on the use of CDM and JI credits. This embracing of project credits will significantly increase the flexibility facilities have in meeting their reduction targets. In addition, Phase 2 includes the use of banking to increase flexibility across time by allowing banked allowances to be used in Phase 3. Each of these market mechanisms is projected to reduce both the EU's Kyoto compliance costs and allowance price volatility. These flexibility mechanisms will be extended into Phase 3 with modifications.

Unfortunately, Phase 1 experience with the ETS did not provide much useful information on the value of market mechanisms or financial instruments in reducing costs or price volatility. The combination of poor emissions inventories, non-use of project credits, and time-limited allowances with effectively no banking resulted in extreme price volatility in Spring 2006, and virtually worthless allowances by mid-2007. The real test for the mechanisms employed by the ETS to create a stable allowance market is Phase 2. Initial indications are that a mature market for allowances appears to be developing, although, like most commodities markets, the allowance market can still be volatile at times. Phase 3 is introducing two new mechanisms in the ETS to further respond to volatility not based on market fundamentals. However, the actual effectiveness of these mechanisms is yet to be proven.

Like the ETS, US cap-and-trade proposals would employ a combination of devices to create a stable allowance market and encourage flexible, cost-effective compliance strategies by participating entities. All include banking. All include use of offsets, although some would place substantial restrictions on their use. Some proposals have incorporated a "safety valve" that would effectively place a ceiling on allowance prices, while others would create a Carbon Market Efficiency Board to observe the allowance market and implement cost-relief measures if necessary. Finally, some incorporate strategic reserves auctions, similar in concept to the EU forward auctioning mechanism, to increase allowance supply without busting the emission cap.

Some see this as a more flexible response with the potential for avoiding or mitigating the environmental impacts of a safety valve (i.e., increased emissions).

Additionally, concern has been expressed in the United States about the regulation of allowance markets and instruments. Based on experience with the ETS, the potential for speculation and manipulation could extend beyond the emission markets. Analysis of ETS allowance prices during Phase 1 suggests the most important variables in determining allowance price changes were oil and natural gas price changes.[69] This apparent linkage between allowance price changes and price changes in two commodities markets raises the possibility of market manipulation, particularly with the inclusion of financial instruments such as options and futures contracts. The concern is sufficient for the Directive to require the EC to examine the situation and the current protections against such activities. Congress may ultimately be asked to consider whether the Securities and Exchange Commission, Federal Energy Regulatory Commission, the Commodities Futures Trading Commission, or other body should have enhanced regulatory and oversight authority over such instruments.[70]

End Notes

[1] Readers unfamiliar with the workings of the European Union may want to read CRS Report RS21372, *The European Union: Questions and Answers*, by Kristin Archick and Derek E. Mix.

[2] Six gases are included under the Kyoto Protocol: carbon dioxide, methane, nitrous oxide, hydrofluorocarbons, perfluorocarbons, and sulfur hexafluoride. The United States has not ratified the Kyoto Protocol and, therefore, is not covered by its provisions. For more information on the Kyoto Protocol, see CRS Report RL33826, *Climate Change: The Kyoto Protocol, Bali "Action Plan," and International Actions*, by Jane A. Leggett.

[3] Norway, a non-EU country, also has instituted a CO2 trading system which is currently linked with the EU-ETS. Various other countries and a state-sponsored regional initiative located in the northeastern United States involving several states have developed a mandatory cap-and-trade system programs. For a review of these emerging programs, along with other voluntary efforts, see CRS Report RL33812, *Climate Change: Action by States to Address Greenhouse Gas Emissions*, by Jonathan L. Ramseur.

[4] P.L. 101-549, Title IV (November 15, 1990).

[5] For further background on the ETS, see CRS Report RL34150, *Climate Change and the EU Emissions Trading Scheme (ETS): Kyoto and Beyond*, by Larry Parker.

[6] More information, including relevant directives, on the EU-ETS is available on the European Union's website at http://europa.eu.int/scadplus/leg/en/lvb/l28012.htm.

[7] Directive 2003/87/EC of the European Parliament and of the Council of 13 October 2003 establishing a scheme for greenhouse gas emissions allowance trading within the Community and amending Council Directive 96/61/EC.

[8] Commission of the European Communities, *Communication from the Commission: Progress towards Achieving the Kyoto Objectives* (November 19, 2008). Other reductions are to be achieve through regulatory measures, such as a CO2 emissions standard for automobiles.

[9] A. Denny Ellerman and Barbara K. Buchner, "The European Union Emissions Trading Scheme: Origins, Allocations, and Early Results," *Environmental Economics and Policy* (Winter 2007), pp. 69-70; and International Emissions Trading Association, "IETA Position Paper on EU ETS Marking Functioning," (no date), p. 3.

[10] For more on the effect of the ETS on Kyoto mechanisms, see A. Denny Ellerman and Barbara K. Buchner, "The European Union Emissions Trading Scheme: Origins, Allocations, and Early Results," *Environmental Economics and Policy* (Winter 2007), p. 84; and International Emissions Trading Association, "IETA Position Paper on EU ETS Market Functioning" (no date), p. 2. For more information on the Kyoto Protocol mechanisms, see CRS Report RL33826, *Climate Change: The Kyoto Protocol, Bali "Action Plan," and International Actions*, by Jane A. Leggett.

11 A Denny Ellerman and Barbara K. Buchner, "The European Union Emissions Trading Scheme: Origins, Allocations, and Early Results," *Environmental Economics and Policy* (Winter 2007), p. 84.

12 European Commission, Directorate General for Environment, *Review of EU Emissions Trading Scheme: Survey Highlights*, (November 2005), pp. 5-7.

13 For a further discussion of Phase 2 implementation issues, see CRS Report RL34150, *Climate Change and the EU Emissions Trading Scheme (ETS): Kyoto and Beyond*, by Larry Parker.

14 For more information, see CRS Report RL33970, *Greenhouse Gas Emission Drivers: Population, Economic Development and Growth, and Energy Use*, by John Blodgett and Larry Parker.

15 On the role of modeling in the first phase, see A Denny Ellerman and Barbara K. Buchner, "The European Union Emissions Trading Scheme: Origins, Allocations, and Early Results," 1 *Environmental Economics and Policy* 1 (Winter 2007), pp. 72-73.

16 Regina Betz and Misato Sato, "Emissions Trading: Lessons Learnt from the 1st Phase of the EU ETS and Prospects for the 2nd Phase," 6 *Climate Policy* (2006), p. 354.

17 For a further discussion, see CRS Report RL33581, *Climate Change: The European Union's Emissions Trading System (EU-ETS)*, by Larry Parker.

18 A Denny Ellerman and Barbara K. Buchner, "The European Union Emissions Trading Scheme: Origins, Allocations, and Early Results," 1 *Environmental Economics and Policy* 1 (Winter 2007), p. 83.

19 European Commission, *Communication from the Commission to the Council and to the European Parliament on the assessment of national allocation plans for the allocation of greenhouse gas emission allowances in the second period of the EU Emissions Trading Scheme,* COM(2006) 725 final (November 29, 2006), p. 11.

20 For a further discussion, see Joseph Kruger, Wallace E. Oates, and William A. Pizer, "Decentralization in the EU Emissions Trading Scheme and Lessons for Global Policy, 1 *Environmental Economics and Policy* 1 (Winter 2007), p. 126; and, Frank J. Convery and Luke Redmond, "Market and Price Development in the European Union Emissions Trading Scheme, 1 *Environmental Economics and Policy* 1 (Winter 2007), pp. 96-7, 107.

21 Data for Phase 2 allowances with a settlement date of December, 2009.

22 International Emissions Trading Association, "IETA Position Paper on EU ETS Market Functioning," (no date), p. 2.

23 Once the EC has approved a country's NAP, including the total number of allowances and the allocation to each covered entity, the allocations can not be re-visited. Attempts to include provisions permitting such post-approval adjustments to a facility's allocation have been uniformly rejected by the EC.

24 European Commission, *Communication from the Commission to the Council and to the European Parliament on the assessment of national allocation plans for the allocation of greenhouse gas emission allowances in the second period of the EU Emissions Trading Scheme,* COM(2006) 725 final (November 29, 2006), p 8; and, A Denny Ellerman and Barbara K. Buchner, "The European Union Emissions Trading Scheme: Origins, Allocations, and Early Results," 1 *Environmental Economics and Policy* 1 (Winter 2007), p. 71.

25 On the mixed record of the EU-ETS and the need for allowance scarcity to a functioning emissions market, see Eric Haymann, *EU Emission Trading: Allocation Battles Intensifying*, Deutsche Bank Research (March 6, 2007). For a generally positive view of ETS market development, see Frank J. Convery and Luke Redmond, "Market and Price Development in the European Union Emissions Trading Scheme, 1 *Environmental*

Economics and Policy 1 (Winter 2007), pp. 97-106. For a more negative view, see Karsten Neuhoff, Federico Ferrario, Michael Grubb, Etienne Gabel, and Kim Keats, "Emissions Projections 2008-2012 Versus NAPs II," 6 *Climate Policy* 5 (2006), pp. 395-410.

[26] Actual 2008 data published by the European Environment Agency in 2010 estimated that the EU-15 were 6.9% below its Kyoto baseline. See European Environment Agency, *Annual European Union Greenhouse Gas Inventory 1990-2008 and Inventory Report 2010*. EEA Technical Report No. 6/2010 (2010), p. 10.

[27] European Environment Agency, *Greenhouse Gas Emissions Trends and Projections in Europe 2009*, (2009) p. 11.

[28] Cyprus and Malta are not Annex 1 countries.

[29] European Environment Agency, *Greenhouse Gas Emission Trends and Projections in Europe 2009* (2009), p. 12.

[30] European Environment Agency, *Annual European Union Greenhouse Gas Inventory 1990-2008 and Inventory Report 2010*. EEA Technical Report No. 6/2010 (2010), p. 10.

[31] European Commission, *Commission Decision of 9 July 2010 on the Community-wide quantity of allowances to be issued under the EU Emission Trading Scheme for 2013*, C(2010) 4658 final, (July 7, 2010).

[32] European Commission, Directive 2009/29/EC *of the European Parliament and of the Council of 23 April 2009 amending Directive 2003/87/EC so as to improve and extend the greenhouse gas emission allowance trading system of the Community* (Brussels, April 23, 2009). Hereinafter referred to as the Directive.

[33] European Commission, *Limiting Global Change to 2 degrees Celsius: The Way Ahead for 2020 and Beyond* (Brussels, January 10, 2007), p. 23.

[34] Joachim Schleich, Regina Betz, and Karoline Rogge, *EU Emissions Trading—Better Job Second Time Around?* Fraunhofer Institute System and Innovation Research (February 2007), p. 23.

[35] Commission of the European Communities, Directive 2003/87/EC, available at http://eur-lex.europa.eu/LexUriServ/ LexUriServ.do?uri=OJ:L:2003:275:0032:0046:EN:PDF.

[36] European Commission, *Proposal for a Directive of the European Parliament and of the Council amending Directive 2003/87/EC so as to improve and extend the greenhouse gas emission allowance trading system of the Community* (Brussels, January 23, 2008).

[37] European Commission, *Directive 2009/29/EC of the European Parliament and of the Council amending Directive 2003/87/EC so as to improve and extend the greenhouse gas emission allowance trading system of the Community* (Brussels, April 23, 2009).

[38] For a further discussion of auctioning and the ETS, see Cameron Hepburn, *et. al*., "Auctioning of EU ETS phase II allowances: how and why?" 6 *Climate Policy* (2006), pp. 137-160.

[39] A Denny Ellerman and Barbara K. Buchner, "The European Union Emissions Trading Scheme: Origins, Allocations, and Early Results," 1 *Environmental Economics and Policy* 1 (Winter 2007), p. 73.

[40] Joseph Kruger, Wallace E. Oates, and William A. Pizer, "Decentralization in the EU Emissions Trading Scheme and Lessons for Global Policy," 1 *Environmental Economics and Policy* 1 (Winter 2007), p. 114.

[41] E.g., Deutsche Bank Research, *EU Emission Trading: Allocation Battles Intensifying*, (March 6, 2007) pp. 2-3; and Regina Betz and Misato Sato, "Emissions Trading: Lessons Learnt from the 1st Phase of the EU ETS and Prospects for the 2nd Phase," 6 *Climate Policy* (2006), p. 353.

[42] A Denny Ellerman and Barbara K. Buchner, "The European Union Emissions Trading Scheme: Origins, Allocations, and Early Results," 1 *Environmental Economics and Policy* 1 (Winter 2007), p. 85.

[43] For more information, see CRS Report RL34150, *Climate Change and the EU Emissions Trading Scheme (ETS): Kyoto and Beyond*, by Larry Parker.

[44] Martina Priebe, *Distributional Effect of Carbon-allowance Trading* (Cambridge, January 12, 2007). Also, see Eurochambres, *Review of the EU* Emission Trading System (June 2007), p. 5.

[45] See Position of the European Parliament adopted at the first reading on 17 December 2008 with a view to the adoption of Directive 2009/.../EC of the European Parliament and of the Council amending Directive 2003/87/EC so as to improve and extend the greenhouse gas emission allowance trading system of the Community (December 17, 2008).

[46] European Commission, Commission Regulation No 1031/2010 on the timing, administration and other aspects of auctioning of greenhouse gas emission allowances pursuant to Directive 2003/87/EC of the European Parliament and of the Council establishing a scheme for greenhouse gas emission allowances trading within the Community. (November 12, 2010)

[47] For example, the US acid rain program provides no allocation of allowances to new entrants; instead, an EPA sanctioned auction is held annually to ensure that allowances are available to new entrants. New entrants can also obtain allowances from existing sources willing to sell them, either directly, through the EPA auction, or via a broker.

[48] A Denny Ellerman and Barbara K. Buchner, "The European Union Emissions Trading Scheme: Origins, Allocations, and Early Results," 1 *Environmental Economics and Policy* 1 (Winter 2007), p. 75.

[49] Karoline Rogge, Joachim Schleich, and Regina Betz, *An Early Assessment of National Allocation Plans for Phase 2 of EU Emission Trading*, Fraunhofer Institute System and Innovation Research (January 2006).

[50] European Commission. *Draft Commission Decision determining transitional Union-wide rules for the harmonised free allocation of emission allowances pursuant to Article 10a of Directive 2003/87/EC* (December 15, 2010).

[51] For more information on climate change and competitiveness issues, see CRS Report R40914, *Climate Change: EU and Proposed US Approaches to Carbon Leakage and WTO Implications*, by Larry Parker and Jeanne J. Grimmett.

[52] Julia Reinaud, *Issues Behind Competitiveness and Carbon Leakage: Focus on Heavy Industry* (October 2008), p. 6.

[53] Carbon Trust, *EU ETS Impacts on Profitability and Trade* (January 2008), p. 4; and The Climate Group, *The Effects of EU Climate Legislation on Business Competitiveness; A Survey and Analysis* (September 2009), p. 8.

[54] European Commission, *Commission Decision of 24 December 2009 determining, pursuant to Directive 2003/87/EC of the European Parliament and of the Council, a list of sectors and subsectors which are deemed to be exposed to a significant risk of carbon leakage* (Brussels, 2009).

[55] European Commission, *Draft Commission Decision of 18 September2009 determining, pursuant to Directive 2003/87/EC of the European Parliament and of the Council, a list of sectors and subsectors which are deemed to be exposed to a significant risk of carbon leakage* (Brussels, 2009).

[56] European Commission, *Emissions trading: Commission welcomes vote to ban certain industrial gas credits* (January 21, 2011).

[57] European Commission, *Proposal for a Directive of the European Parliament and of the Council amending Directive 2003/87/EC so as to include aviation activities in the scheme for greenhouse gas emission allowance trading within the Community* (Brussels, December 12, 2006).

[58] European Commission, *Proposal for a Directive of the European Parliament and of the Council amending Directive 2003/87/EC so as to improve and extend the greenhouse gas emission allowance trading system of the Community* (Brussels, January 23, 2008), p. 4.

[59] As stated by CRS in 1992: "For an economic incentive system to be effective, several preconditions are necessary. Perhaps the most important is data about the emissions being controlled. Such data are important to levy any tax, allocate any permits, and enforce any limit." CRS Issue Brief IB92125, *Global Climate: Proposed Economic Mechanisms for Reducing CO2*, by Larry Parker (archived November 16, 1994), p. 9.

[60] Section 821, *1990 Clean Air Act Amendments* (P.L. 101-549, 42 USC 7651k).

[61] S.Rept. 101-952.

[62] See 40 CFR 75.13, along with appendix G (for CEMs specifications) and appendix F (for fuel analysis specifications.

[63] P.L. 110-161

[64] 40 CFR Part 98 (effective December 29, 2009).

[65] For more background, see CRS Report RL33581, *Climate Change: The European Union's Emissions Trading System (EU-ETS)*, by Larry Parker.

[66] Ibid., p 3.

[67] Unlike a carbon tax which uses the existing currency system to control emissions—be it euro or dollars.

[68] Deutsche Bank Research, *EU Emission Trading: Allocation Battles Intensifying* (March 6, 2007), p. 2.

[69] Maria Mansanet-Bataller, Angel Pardo, and Enric Valor, "CO2 Prices, Energy and Weather," 28 *The Energy Journal* 3 (2007), pp. 73-92.

[70] For a discussion of regulation of allowances as a commodity and implications for a greenhouse gas emissions market, see CRS Report RL34488, *Regulating a Carbon Market: Issues Raised By the European Carbon and US Sulfur Dioxide Allowance Markets*, by Mark Jickling and Larry Parker.

In: The European Union's Emissions … ISBN: 978-1-62081-955-5
Editors: A.R. Moore and M.F. Ryan

Chapter 2

AVIATION AND THE EUROPEAN UNION'S EMISSION TRADING SCHEME*

Jane A. Leggett, Bart Elias and Daniel T. Shedd

SUMMARY

Beginning January 1, 2012, most carbon dioxide (CO_2) emissions from commercial flights to, from, and within the European Union (EU) are covered by the EU Emission Trading Scheme (EU ETS). Flights are covered regardless of whether the airline or operator is based in the EU region. The EU ETS caps aviation emissions of CO_2 in 2012 at 97% of the average in 2004-2006 and at 95% in each year from 2013 to 2020. Each April, beginning 2013, covered aircraft operators must turn in emission "allowances" (permits) equal to the previous year's emissions from their flights arriving at or departing from EU airports. Airline operators will receive free allowances for 82%- 85% of their 2010 emissions. Airlines that have more allowances than they need may sell them to others or save them for future use. Airlines that need more allowances may buy them from EU auctions, other carriers, other emission sources in the EU ETS, brokers, or international emission trading mechanisms. A small reserve of free allowances will be available for new or rapidly expanding airlines.

* This is an edited, reformatted and augmented version of a Congressional Research Service publication, CRS Report for Congress R42392, from www.crs.gov, prepared for Members and Committees of Congress, dated March 7, 2012.

The entry into force of an EU law covering international aviation emissions is a significant move in a two-decade process concerning whether and how aviation emissions of CO_2 may be abated. Even among those who agree on the benefits of reducing greenhouse gas (GHG) emissions, how to share reductions across nations and sectors remains a continuing, major controversy internationally and within countries. Emissions from international air transport have, since early negotiations, posed a particular challenge.

The 1997 Kyoto Protocol (to which the United States is not a Party) specified that Parties should pursue limitation or reduction of GHG emissions from aviation fuels, working through the International Civil Aviation Organization (ICAO), an agency of the United Nations. In 2008, the EU cited a lack of "substantive progress" in ICAO and legislated to include aviation in its existing EU ETS by 2012. ICAO members have agreed to a variety of voluntary actions and goals. In October 2011, faced with impending inclusion of aviation emissions in the EU ETS, the ICAO Council agreed to accelerate its work, including continuing to explore market-based measures, CO_2 standards for new aircraft, and other options.

EU officials have stated their preference for agreeing on global, binding measures in ICAO. One official stated that the EU would agree to suspend inclusion of aviation in the ETS only if a new global ICAO scheme met three conditions: it must deliver more emissions reductions than the EU ETS on its own; it must have targets and measures; and any action must be non-discriminatory and apply to all airlines. Alternatively, EU law allows exemptions for incoming flights from countries that have adopted "equivalent measures" to reduce emissions. Some countries are likely negotiating with the EU for an equivalent measures exemption for their airlines.

Air carriers from the United States and other countries have vociferously objected to inclusion of international aviation in the EU ETS. The US government and other nations have pressed the EU to exclude foreign carriers. Two bills in Congress address the controversy. The US House of Representatives passed H.R. 2594, the European Union Emissions Trading Scheme Prohibition Act of 2011. It would prohibit US aircraft operators from participating in the EU ETS. It also would direct the Administration to negotiate and take other actions to ensure that US civil aircraft operators are not penalized by any unilateral EU regulation of GHG emissions. A similar bill with more flexibility, S. 1956, was introduced in the Senate. In January 2012, House and Senate conferees on reauthorization of the Federal Aviation Administration (FAA) agreed to a sense of the Congress resolution opposing the EU action.

INTRODUCTION

Absent regulation, greenhouse gas (GHG)[1] emissions from aviation are projected to rise rapidly due to the continuing increase in the number of passenger and cargo flights. To address these emissions, the European Union (EU) has chosen to regulate aviation emissions in its regional Emissions Trading Scheme (EU ETS). Aircraft operators flying to, from, or within the EU are covered whether or not they are based in the EU. Many aircraft operators oppose being regulated, and many countries have objected that the system infringes on their sovereignty.

The US government is among those disputing the EU's inclusion of international aviation in the EU ETS, contending that any policies should be agreed under the International Civil Aviation Organization (ICAO) or bilaterally. US airlines are pressing the Congress for action to prevent the EU ETS from applying to them, and Members have agreed to a sense of the Congress resolution objecting to the scheme.

This report provides background on inclusion of international aviation in the EU ETS and sets forth possible options for congressional consideration. It summarizes the broad agreements internationally to address GHG emissions, and explains why emissions from aviation are a particularly difficult problem. The report then describes the EU's approach to abating GHG emissions, principally through its EU ETS, and how the scheme specifically covers aviation.

The report outlines reasons that many international airlines and non-EU nations oppose the EU regulation, including a review of potential financial, technological, and competitiveness issues. It summarizes December 2011 rulings by the European Court of Justice on litigation raised by US airlines. Responsive actions by the US government and other countries are identified. Finally, the report offers options for congressional action as well as conclusions. Appendixes identify key upcoming dates in the EU ETS coverage of aviation, and summarize related activities under ICAO.

A Two-Decade Process to Address Aviation Emissions

EU inclusion of aviation in the EU ETS is the latest step in a two-decade process concerning whether and how aviation emissions of GHG may be abated. Formal international agreement to abate GHG emissions extends back to 1992, when 193 countries plus the European Union agreed to the United

Nations Framework Convention on Climate Change (UNFCCC). In the 1997 Kyoto Protocol, the industrialized ("Annex I") Parties agreed to reduce their GHG emissions by, on average, 5% below 1990 levels during the period 2008 to 2012. The United States is a Party[2] to the UNFCCC but did not accede to the Kyoto Protocol.

The legally binding commitments under the Kyoto Protocol were understood to be a first step toward halting the rise of atmospheric GHG concentrations and slowing human-induced climate change. Parties have enacted a variety of measures aimed at abating GHG emissions, but most countries have expressed alarm that existing measures would be insufficient to stabilize human-induced climate change.

In 2010, the UNFCCC Parties adopted a global goal of holding a GHG-induced increase in the Earth's temperature below 2° Celsius (3.6° Fahrenheit).[3] Achieving this goal would likely require more than halving current global GHG emissions.[4]

Even among those who agree on the benefits of reducing GHG emissions, how to share reductions across nations and sectors of the economy remains a continuing, major controversy internationally and within countries.[5] Emissions from international air transport have, since early negotiations, posed a particular challenge.[6]

Under UNFCCC rules, almost all emissions are counted in the nation where they occur. Much of the emissions associated with international aviation, however, occurs in international air space. As with other sectors, aviation emissions could be counted in, and regulated by, the country where the aviation fuel is sold; however, small countries (such as the Netherlands, Malta, and Singapore) for whom sales of aviation and marine bunker fuels represent an unusually large portion of the economy have objected to this option.

In part, they argue that they would have little influence over the technologies used by carriers based elsewhere. They also argue that, if all countries selling aviation fuels did not regulate emissions, fuel sales would shift to uncontrolled locations. This could result in economic transfers, losses, and emissions "leakage"—not the intended GHG reductions. As one of a package of compromises, the 1997 Kyoto Protocol (to which the United States is not a Party) specified that the industrialized country Parties should pursue limitation or reduction of GHG emissions from aviation bunker fuels, working through ICAO.[7]

The 1944 Chicago Convention on International Civil Aviation established the Provisional International Civil Aviation Organization, which became the

ICAO in 1947.[8] Since 1947, ICAO has been an entity of the United Nations. ICAO supports cooperation among its member countries on non-binding standards and recommended practices for safety, security, environmental protection, and other matters affecting civil aviation. The United States is a Signatory to the Chicago Convention and is one of 190 current member States of ICAO. While the European Union (EU) is not a Signatory of the Chicago Convention, all of its 27 member States are. The EU maintains an observer status within ICAO.

Since the 1997 mandate in the Kyoto Protocol, ICAO has produced technical information and a variety of voluntary options and recommendations pertaining to the limitation of aviation GHG emissions. In 2004, ICAO ruled out the option of a *global* emission trading scheme for aviation. Instead, it produced guidance for member States should they include international aviation in their own emission trading schemes. ICAO urged that inclusion be with "mutual agreement" and non-discriminatory. For several years, the EU warned that, without stronger action under ICAO, it would include the aviation sector in the exiting EU ETS.

In EU Directive 2008/101/EC, adopted on November 19, 2008, the EU Council and Parliament amended the EU ETS to include aviation sources of CO_2 emissions,[9] effective January 1, 2012. The lack of greater progress under ICAO also led Australia and New Zealand to proceed with including the aviation sector in their domestic carbon tax and GHG cap-and-trade schemes. However, both nations exempted international aviation fuels from their requirements.[10]

In October 2011, faced with impending inclusion of aviation emissions in the EU ETS, the ICAO Council agreed to accelerate its work and continue to explore market-based measures to "reach a global solution." The ICAO Secretary General pledged to produce a proposal by the end of 2012. An EU official has stated that ICAO is "the right place to advance action on market-based measures for aviation and the EU strongly supports its work."[11] However, the official also stated that "what is agreed on market-based measures in ICAO must be a global solution and deliver more emissions reduction than under measures now in place, including by 30 countries applying the EU emissions trading system," and must be agreed in a treaty (i.e., legally binding).[12]

A later section of this report ("Activities in ICAO") and *Appendix B* discuss in more detail the processes under ICAO and interactions with the EU.

View of Airlines for America (A4A)

"The airline industry advocates for a single, global sectoral approach to be managed by ICAO, which stipulates that our industry achieve a 1.5% average annual improvement in carbon and fuel efficiency through 2020, carbon-neutral growth from 2020, subject to critical government infrastructure and technology investments such as air traffic control modernization, and an aspirational 50% reduction in CO_2 emissions by 2050, relative to 2005 levels."

Source: Airlines for America, "ATA[13] Calls EU ETS Application to US Airlines Illegal," Washington, DC, July 5, 2011, available at http://www.airlines.org/ Pages/ATA- Calls-EU-ETS-Application-to-US-Airlines-Illegal.aspx.

WHY ADDRESS AVIATION EMISSIONS?

Aviation contributes to climate change at regional and global scales through a variety of emissions, including carbon dioxide, nitrogen oxides (leading to ozone pollution), water vapor, particulate matter (soot), and other pollutants. Scientists estimate that all these types of emissions from global aviation in 2005 accounted for about 5% of the human effect on global temperature rise.[14] Emissions of carbon dioxide (CO_2) are about half of aviation's effect, or about 2%-3% of the total.[15] The EU's Emission Trading Scheme regulates only CO_2 emissions from aviation.

Aviation: Relatively Large Among Many Small Emission Categories

Although fossil fuels account for roughly three-quarters of all CO_2 emissions, fossil fuel combustion is by a diffuse set of sources: powerplants, vehicles, home heating, industrial processes, etc. Many countries that are addressing climate change have acted first on the largest sources of emissions: powerplants and motor vehicles. After those source categories, the remaining categories typically are small shares of the total. Of course, as the emissions of the controlled sources shrink, the shares of the remaining categories grow. Aviation also combusts fuel, and its share of CO_2 emissions is about 2%-3% of the global, human-related total (excluding changes in forests and other land uses).

This is about the same as all CO_2 emissions from Germany or Korea. International aviation emissions—only from flights that cross sovereign boundaries—are a smaller share—about 1.5% in 2009 (about the same as Italy's or Mexico's CO_2 emissions).[16]

One of the Fastest Growing Sources of Emissions

Aviation has been one of the most rapidly growing sources of CO_2 emissions. *Figure 1* illustrates the indexed changes in CO_2 emissions from international aviation compared to all fossil fuel use, for the EU and the world. In the EU, fossil fuel emissions fell by about 5% from 1990 to 2009, while the emissions from aviation fuels sold in the EU rose by about 80%. This was more than the increase of all emissions from Spain.

Globally, also, aviation emissions rose faster than all fossil fuel emissions, though the difference was less. ICAO estimated that CO_2 emissions from all aviation (not just international) almost doubled from 1990 to 2006. This compared with a 34% increase from all fossil fuel use.[17]

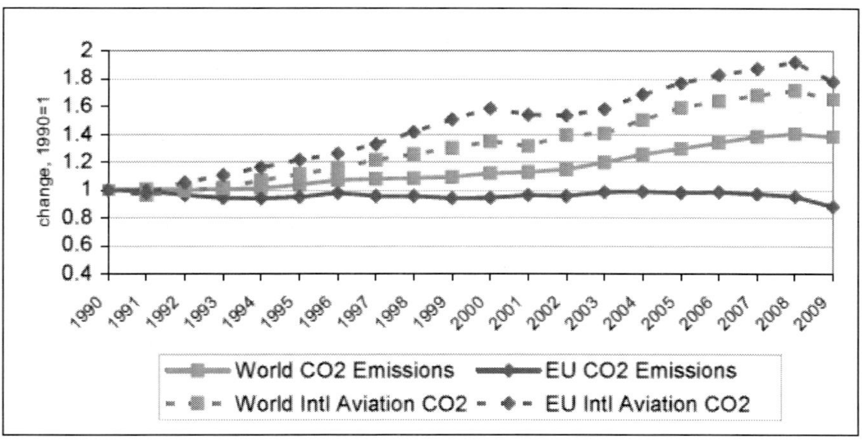

Source: CRS figure using data from IEA CO_2 emissions estimates, 1971-2009. Data base from 2011.

Figure 1. Percent Changes in CO_2 Emissions from International Aviation and All Fossil Fuel Use, in the World and in the EU.

Counteracting Influences: Demand Growth and Efficiency Gains

Past and projected emissions growth in the aviation sector are influenced by two counteracting forces: expanding aviation services and improving efficiency.

Expansion of Air Transport Demand

Carbon dioxide emissions from aircraft are increasing due to expanding demand for aviation services. Economic growth, globalization, falling air fares (after accounting for inflation), liberalization of air service markets, and the desire for rapid transport have driven demand for passenger travel and cargo transport. Both passenger and cargo traffic have grown much more rapidly than economic growth over past decades. From 1999 to 2009, carriers saw a 4.3% average annual growth in revenue-passenger-kilometers (RPK), while cargo shipments saw a 2.6% annual average growth despite economic weakness at the end of the decade.[18]

Past Improvements in Fuel Efficiency and Emissions per Passenger-Mile

Aviation emissions would be higher if not for improved fuel efficiency in commercial aviation industry. US domestic air operations reduced their fuel intensity (energy consumed per passenger-mile carried) by 41% between 1990 and 2008—more than any other mode of passenger transportation—and US airlines' international air passenger operations reduced fuel intensity by 24% (See *Figure 2*).[19] More recently, the US carriers' trade group reports that carriers have done a better job of keeping pace with growth, reducing absolute fuel burn and emissions by 10% from 2000 to 2010.[20]

Further Improvements May Help Suppress Emissions

Several changes are projected to reduce significantly the rates of aircraft fuel consumption and emissions over the next decade. More efficient aircraft designs will be coupled with air traffic management improvements to reduce delays and minimize inefficient routing of aircraft.

Alternative fuels, including biomass-based fuels, could also penetrate the aviation market with benefits for emissions and energy security.

Improved air traffic management will likely help suppress emissions and save fuel. For example, in the United States, the FAA estimates that its Next Generation Air Transportation System (NextGen) air traffic modernization efforts will reduce aviation fuel consumption by an estimated 1.4 billion gallons, cutting cumulative CO_2 emissions by 14 million tons between 2010 and 2018.[21] Similarly, the EU has set goals for its Single European Skies Air Traffic Management Research Initiative (SESAR) to reduce per flight fuel consumption and CO_2 emissions by 10% from improvements in air traffic management alone by 2020.[22]

1990-2008, 1990 average Btu/passenger-mile = 1

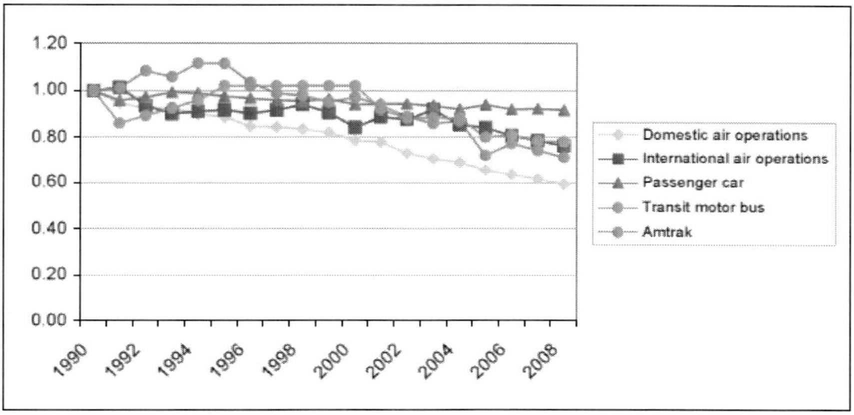

Source: CRS figure using data from Bureau of Transportation Statistics. "National Transportation Statistics." US Department of Transportation, Table 4-20. Data extracted January 23, 2012.

Figure 2. Change in Fuel Intensity of US Passenger Transportation by Mode.

Additional CO_2 emissions reductions may be realized from a gradual transition to more efficient aircraft fleets. For example, Boeing Commercial Airplanes claims that its newly certified wide-body B-787 Dreamliner, the first large airliner to make extensive use of lightweight composite materials in its construction, uses 20% less fuel and produces 20% less CO_2 than the similarly sized Boeing B-767.[23] Similarly, Airbus touts that its A350 airplane currently in development will achieve fuel burn and CO_2 emissions reductions of 25% compared to its current long-range competitor through use of a lighter composite fuselage, improved engines, and an advanced wing design optimized for cruise flight.[24] As airlines gradually phase out older aircraft and replace them with newer, more fuel efficient models, corresponding reductions in CO_2 emissions may be achieved. The advanced concepts in the B-787 and A350 will largely be limited to medium- and long-range operations. In the near term, short haul aircraft like the B-737 and A320 are expected to get more fuel efficient engines and some improvements in aerodynamics, but these changes are likely to yield much smaller comparative reductions in CO_2 emissions than technologies for newly designed aircraft. Observers expect that deliveries of A320s and B-737s will continue at least through 2020.[25]

Biomass-based fuels can have net CO_2 emissions approaching zero, when they are produced with feedstocks (i.e., plant material) that sustainably gets carbon from the atmosphere (i.e., through photosynthesis). US airlines have used biomass aviation fuels in commercial flights. However, the cost to produce these fuels remains uncompetitive, with future supply uncertain. Some in the industry are bullish on biofuel potential: reportedly, the CEO of Alaska Air Group affirmed, "What we need is an adequate, affordable and sustainable supply.

To the biofuels industry: If you build it, we will buy it."[26] The International Air Travel Association has said, "Savings of up to 80% [of carbon emissions] are on the table if the industry can get it right."[27] Others caution that biofuels may not be commercially viable for years, and supply may be too limited to reduce emissions.[28]

Aviation Emissions Projected to Grow Rapidly

Further improvements to aircraft design, air traffic operations, and commercial biofuels are anticipated to yield meaningful reductions in aircraft fuel efficiency and CO_2 emissions per ton-kilometer. Nonetheless, a net increase in fuel consumption and CO_2 emissions is anticipated from continued growth in worldwide flight operations. ICAO forecasts that CO_2 emissions from aviation are likely to grow by 3%-4% annually for the next few decades.[29]

More specifically, ICAO projects that aviation's CO_2 emissions would grow globally to 2.5 to 2.9 times the 2006 emissions by 2036, and 3.5 to 4.5 times the 2006 emissions by 2050.[30] (ICAO projections do not include possible penetration of non-fossil fuel use into the aviation market.) Under ICAO's scenarios of future aviation emissions, the sector's emissions in 2050 would comprise about 20% of the global emissions compatible with holding projected global temperature increases to 2°C.[31]

As the EU and other countries implement their laws to reduce other major sources of GHG emissions, aviation emissions would continue to increase in absolute terms and as a share of total GHG emissions. Hence, the EU has chosen to address aviation emissions in its comprehensive climate change policy package.

EU's Legal Obligations to Reduce GHG Emissions

Under the Kyoto Protocol, the European Union (EU) agreed to a GHG target of 8% below 1990 levels on average in the period from 2008 to 2012.[32] In the context of international negotiations on how to mitigate GHG beyond 2012, the EU committed to reduce its GHG emissions to 20% below 1990 emissions by 2020.[33] The EU's overall GHG targets have been embodied in EU law[34] with corresponding measures enacted to achieve the emission reduction goals. The EU and its member States must take the necessary measures to comply with the emissions levels embodied in EU law (Decision 2002/358/EC).

In December 2011, the EU and other Parties agreed to a second commitment period of the Kyoto Protocol, beginning in 2013. While the terms are not yet settled, all Parties expect that the legally binding GHG target for the EU will be at least as stringent as the 20% target embedded in EU law. For 2050, the EU has adopted a strategy proposed to reduce GHG emissions to 80%-85% below 1990 levels.

In the GHG accounting rules under the UNFCCC, emissions from international aviation and maritime transport ("bunker fuels") are not tallied in each Party's inventory subject to its legally binding GHG reduction target.[35] Still, the language of the Kyoto Protocol mandates the Parties to pursue limitation or reduction of aviation emissions.

The EU enacted its Emissions Trading Scheme, a "Climate and Energy Package," and other measures in a community-wide approach to meeting its GHG targets. An "Effort Sharing Decision" set national emissions targets for 2020 for each member State for sectors not covered by the EU ETS, with an overall emissions reduction from non-ETS sectors of 10% from 2005 levels. The emissions from most sectors are now regulated by the EU or its member States, either by performance standards (e.g., for motor vehicles) or by being included in the EU's Emissions Trading Scheme (EU ETS). The *Text Box* shows targets for some major sectors, in and outside of the EU ETS. The EU is also expanding incentives for renewable energy and pursuing carbon capture and storage technologies.

The EU's GHG Emissions Targets by Sector

The EU has enacted a set of nearly comprehensive policies to control GHG emissions across most sectors.

The EU ETS is only one piece of the strategy. Across all programs, the GHG targets for 2020 for aviation and other sectors, compared to 2005 levels, are:

- Aviation: 5% below 2004-2006 average levels
- Electricity generation and industry: 21% below 2005 levels in the EU ETS
- Buildings: 10% below 2005 levels
- Agriculture: 10% below 2005 levels
- Non-aviation transport: 10% below 2005 levels

The EU's Emissions Trading Scheme covers electricity generation and energy intensive industry, constituting about 50% of all EU GHG emissions. For the sectors other than aviation, the emission cap by 2020 will be 21% below 2005 emission levels. The EU's policy also requires by 2020 that renewable energy provide 20% of overall energy consumption, and that energy efficiency improve by 20%.

HOW THE EU'S EMISSIONS TRADING SCHEME WORKS

The "Emissions Trading Scheme" (EU ETS) is one of the primary instruments the EU uses to achieve its legislated GHG reduction targets. The EU ETS is a "cap and trade" program.[36] Under the EU ETS, the European Union (EU) plus Iceland, Lichtenstein, and Norway have set a cap, or a maximum quantity, on annual carbon dioxide (CO_2) emissions from major industrial sources. Each individual source is given or must purchase a quantity of "allowances" or permits to emit one metric ton of CO_2. 37 At the end of each year, each source of emissions must turn in a number of allowances at least equal to its emissions for the year. If the source does not turn in sufficient allowances, it must pay a fine. Over time, the emissions caps will decline and fewer allowances will be issued, forcing CO_2 emissions to decline as well to the enacted levels of the cap. Eventually, the EU foresees auctioning all allowances rather than giving a portion free to emission sources. (How aviation is included in the EU ETS is discussed in a later section; see "Aviation Provisions in the EU ETS.")

To minimize compliance costs and give flexibility to emissions sources, the EU ETS allows emission allowances to be traded. A source that has more allowances than it needs may sell the extra to another source that needs more, or to other entities, such as an allowance broker. The allowance market sets the price for the allowances.

If a source can reduce one ton of emissions at a cost lower than the price of an allowance, the EU ETS gives it the profit incentive to reduce emissions and sell allowances.

It also gives businesses (e.g., technology companies) motivation to develop and deploy new, lower-cost means to reduce emissions.

Currently, the EU ETS covers aviation in addition to other CO_2-emitting sectors: electricity generation; fuel combustion plants; oil refineries; iron and steel works; and manufacturers of cement, glass, lime, bricks, ceramics, pulp and paper. At present, the EU ETS covers about half of the EU's CO_2 emissions.

Beginning in 2013, the EU ETS will include petrochemicals, ammonia, and aluminum manufacturers, and will extend emissions covered from some sources (not aviation) to include nitrous oxides (N_2O) and perfluorocarbons (PFC).

AVIATION IN THE EU ETS

The EU reports that about 3% of its greenhouse gas emissions come from civil aviation activities. The share coming from flights between EU and non-EU airports is smaller.

The EU is concerned about aircraft emissions because, as *Figure 1* illustrates, from 1990 to 2010, overall CO_2 emissions in the EU declined by about 5% while aviation emissions rose by about 80%. Emissions from aviation have been growing faster than those from any other sector. They are projected to continue rising due to increases in air travel and air-cargo shipments.

Civil aviation emissions account for 11% of GHG emissions covered by the EU ETS. They are the second largest share of any sector, after power generation (see *Figure 3*).

Civil aircraft serving the EU emit more GHG than petroleum refineries and steel plants, and more than additional sectors, such as petrochemicals and aluminum, that must comply beginning in 2013. The EU determined that, to progress towards its goal of stabilizing CO_2 concentrations in the atmosphere, and to treat aviation comparably with other sectors, aviation emissions needed to be regulated.

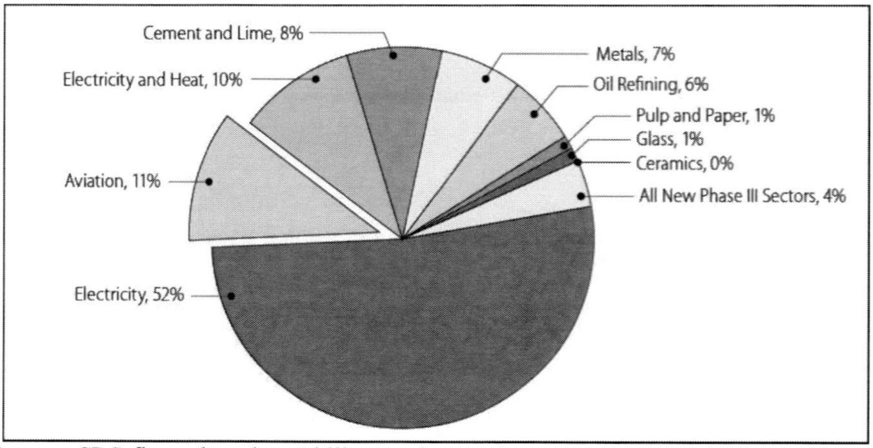

Source: CRS figure based on Philip Good. "Tourism, aviation and emissions trading" DG Climate Action, European Commission. March 14, 2010.

Figure 3. Sectoral Shares of the CO_2 Emissions Cap of the EU ETS in 2013.

The EU has used a multi-pronged approach to address aviation GHG emissions, seeking measures in its member States and multilaterally. According to the EU, the EU ETS is one of several policy instruments it is using to reduce aviation emissions; other instruments include research on new technologies and biomass fuels (the Clean Sky Joint Technology Initiative), work under ICAO to develop a CO_2 standard for new aircraft, and modernization of air traffic management under the Single European Sky and SESAR program (Single European Sky Air Traffic Management Research).[38]

Though the EU's existing aviation programs are expected to help slow the growth of emissions, EU countries have viewed global actions as necessary ultimately to reducing aviation emissions. In particular, the EU has pressed for common standards to be adopted through ICAO for CO_2 emissions for new aircraft, and for mandatory global market-based measures (MBM). MBM could include an emissions fee or emissions trading system.

Since 1998, ICAO Members, including the member States of the EU, have worked through ICAO toward measures to abate aviation emissions. The EU recently stated

> Europe has worked within ICAO and the UNFCCC to reach a global agreement on aviation emissions for over 15 years. However progress has been slow and does not match the urgency of the climate problem. Europe has taken action due to the lack of substantive progress at the international level.[39]

In 2008, the European Parliament and Council enacted a new law, Directive 2008/101/EC, amending the EU ETS (Directive 2003/87/EC) to include aviation activities.[40]

Aviation Provisions in the EU ETS

Effective January 1, 2012, the EU ETS covers most CO_2 emissions from aviation to, from, and within the region. The EU ETS caps aviation emissions in 2012 at 97% of the average 2004-2006 emissions on covered flights. The cap declines to 95% of historical emissions for 2012-2020, and may be further reduced beyond 2020 pending EU review. The requirements apply to all civilian aircraft arriving at or departing from airports in the EU, Norway, Iceland, and Liechtenstein, not just airlines based in those territories.

The EU ETS exempts certain types of flights from the cap-and-trade scheme, including military aviation; search & rescue flights; state flights transporting third countries' heads of state, heads of government, and government ministers; and police flights. Also exempted are operators that meet *de minimis* criteria.[41] Many of the hundreds of US aircraft operators may be excluded from the EU ETS because they fly below the *de minimis* criteria.

For administrative simplicity, each airline will be administered by only one of the participating countries. Aircraft operators have been listed by the EU and advised to contact the "competent authority" in the country that will administer their participation in the scheme. Covered airlines monitored their activities and emissions during 2010 and were required to report them to the competent authority by March 31, 2011. As with other emitters in the EU ETS, each aircraft operator will have to turn in allowances to the "competent authority" to cover its annual emissions by April of the following year.

For each metric ton of CO_2 a source emits, it must turn in one allowance. During 2012, 85% of the emissions allowances to the aviation sector will be given free to aircraft operators and 15% auctioned. During the period 2013 to 2020, 82% will be given free, while 15% will be auctioned and 3% will go into a reserve for rapidly expanding airlines and new entrants into the market.

The initial allocation of free allowances to each operator is based on the "benchmark" times the number of ton-kilometers its aircraft flew on flights to, from, or within the EU during 2010. Each operator receives 0.6797 free allowances (the "benchmark") for each covered 1000 ton-kilometers of covered flights. In 2013 to 2020, when the cap declines to 95% of the 2004-2006 emissions level, the benchmark shrinks as well (see *Table 1).*

Table 1. Emission Allowances for Aviation in the EU ETS (one allowance permits one metric ton of CO_2 emission)

Trading Period	Annual Area-Wide Cap	Distributed Free	Benchmark (per 1000 ton- km)	Auctioned	Special Reserve (total over 8 years)
2012	214,777,670	182,561,019	0.6797	32,216,651	0
Each year 2013-2020	210,349,264	172,486,396	0.6422	31,552,390	50,483,824

Source: CRS table with information from Decision of the EEA Joint Committee, No 93/2011 (July 20, 2011), http://ec.europa.eu/clima/policies/transport/aviation/allowances/docs/eea_20072011_en.pdf.

Note: This table covers all countries participating in the system: the EU plus Norway, Lichtenstein, and Iceland.

An airline that requires more allowances than its free allocation may bid for them at periodic auctions held by national governments or purchase them from sources holding allowances that wish to sell. Because the aviation requirements are part of an "open" emission trading system, these allowances may be traded with allowances from other emission sources under the EU ETS, from brokers, or from approved "certified emission reductions" generated by reducing GHG from projects in developing countries. Similarly, airlines receiving allowances may sell them to others.

If an airline does not comply with the EU ETS, it is subject to an "excess emissions penalty." The penalty is set at €100 (US$130) for each ton of CO_2 emissions exceeding the allowances surrendered—an amount many times the highest expected allowance prices in the market. The aircraft operator also remains liable for submitting allowances in the following year for those emissions. The EU law provides for sanctions that include the possibility that a repeatedly noncomplying airline might, as a last resort, be banned from operating in the EU.[42]

Disposition of Revenues

In each year, 15% of the aviation allowances will be auctioned by member States. While the Directive leaves it to those member States to decide how to use revenues generated by auctioning allowances, it also instructs that revenues

> should be used to tackle climate change in the EU and third countries, inter alia, to reduce greenhouse gas emissions, to adapt to the impacts of climate change in the EU and third countries, especially developing countries, to fund research and development for mitigation and adaptation, including in particular in the fields of aeronautics and air transport, to reduce emissions through low-emission transport and to cover the cost of administering the Community scheme. The proceeds of auctioning should also be used to fund contributions to the Global Energy Efficiency and Renewable Energy Fund, and measures to avoid deforestation.[43]

Member States must inform the European Commission (EC, the EU implementing body) how revenues were used. Though an instruction like this to member States is unusual in EU law, some airline representatives have criticized the discretionary nature of it. Some express concern that the funds may not be used within the aviation sector and/or that they may not be used to reduce GHG emissions.

Potential EU Exemptions with Equivalent Measures by Other States

EU Directive 2008/101 provides for the EC to consider "options available in order to provide for optimal interaction" between the EU ETS aviation provisions and the measures adopted by a "third country" (i.e., non-EU country) to reduce aviation emissions. The options may include exemption from the allowance requirements of flights incoming from a third country that has adopted its own measures to reduce aviation emissions; in this case, outbound flights from the EU to the third country would be subject to the EU ETS, but inbound flights would be subject to the legal regime of the country in which they originate. The exclusion would require an amendment to the EU law. The options may include amendments to other "non-essential elements" of the aviation provisions, as well. Also, the EC may recommend that the EU Council open negotiations with a third country on an agreement, with unspecified terms and conditions.

The EU law directs that the EU and its member States continue to seek agreement on global measures to reduce GHG emissions from aviation. This would include negotiations, for example, under ICAO or the UNFCCC. Recently, an official stated that the EU would agree to suspend inclusion of aviation in the EU ETS only if a new global ICAO scheme met three conditions: it must deliver more emissions reductions than the EU ETS on its own; it must have targets and measures; and any action must be non-discriminatory and apply to all airlines.

Estimated Impacts on US Airlines

The EU Commission has identified hundreds of US aircraft operators that fly to or from EU airports and therefore could be subject to the EU ETS. Many are freight operators that provide relatively infrequent service to the EU, and therefore may fall below the de minimis level of emissions for compliance. Large US-based passenger carriers, such as United Airlines, Delta, and American, have sufficient European operations to be covered by the scheme, as do major express cargo carriers FedEx and UPS. The first distribution of free allowances for 2012 occurred in February 2012. Notably, with more than a year before aviation allowances must be surrendered to comply with the EU ETS, a number of large airline companies have been buying European allowances while prices are low. Several commodity exchanges are launching trading in the EU aviation allowances (EUAAs).

Complying with the EU ETS is likely to increase costs for international air carriers, although the extent and significance are likely to depend on developments that cannot be known at this time. Some of the controversy may be explained by different views concerning:

- emission reduction potentials of technological and operational strategies;
- projected future allowance prices;[44]
- effects on demand for aviation services;
- effects on research and development (R&D) and investment in new systems;
- potential for profit opportunities and windfalls for some; and
- competitive dynamics.

These issues are explained in the sections below.

Emission Reduction Potential of Technical Improvements

Advancing technologies are expected to help aircraft operators to reduce their emissions, and consequently some portion of their needs for allowances. The aviation industry, in conjunction with many governments, is developing more fuel efficient technologies (as discussed in a previous section). Some more efficient technologies would be phased in with new aircraft, while others could be retrofitted, or "dropped in" (e.g., winglets, or certain low carbon biofuels). New air traffic control systems are projected to curb aircraft fuel consumption and emissions over the next decade. The rate at which these operational systems improve will depend on government funding, and will affect private companies' fuel efficiency and CO_2 emissions.

Anticipating these changes, the International Air Transport Association, representing 240 airlines and 84% of air traffic, has adopted a voluntary goal of reducing fuel efficiency and CO_2 emissions per revenue tonne-kilometer by at least 25% by 2020 compared to 2005 levels.

Voluntary efficiency improvements are likely insufficient to assure compliance with the EU ETS. One uncertainty is the rate and cost (or savings) of deploying further efficient technologies and biofuels with and without fiscal incentives or regulatory requirements, such as the EU ETS. The EU considers that the certainty of an emissions cap reduces the risk to businesses that invest in advanced technologies that help reduce emissions.

Different Assumptions about Allowance Prices

A principal reason for differences in cost estimates is varying future prices assumed for carbon allowances. Allowance prices in the coming decade could be higher or lower than recent prices. On the one hand, prices are likely to rise in the future as the European economies strengthen, more sources are capped by the emissions trading system, and the emissions caps tighten. On the other hand, slower than anticipated economic growth, advances in efficiency and technologies in other sectors, and potentially large supplies of certified emission reductions in developing countries could dampen prices for many years.

Two contrasting cost estimates underscore the importance of allowance price assumptions: In 2009, the Air Transport Association of America (ATA)[45] estimated that the EU ETS would cost US carriers \$3.1 billion over nine years, or an average of \$344 million per year.[46] A Bloomberg Government analysis[47] in November 2011 prepared a critical analysis for comparison: It estimated that, depending on the price of allowances, the total compliance cost would range between $0.9 billion and $4.2 billion over nine years, if airlines do not shift to more fuel-efficient planes or to lower-carbon fuels. At average allowance prices since 2005 (US$11.40 per metric ton CO_2), the cost would be $2.1 billion over nine years, or $233 million per year. Bloomberg concluded that the allowance price would need to be 50% higher than the average since 2005 to reach the ATA's cost estimate.[48]

A near-term economic slowdown in the EU could serve to reduce the relative impacts on US operators over the next few years.[49] In the longer term, allowance prices may be tempered by low-cost emissions reductions in other sectors and other regions. As recommended by ICAO, the EU ETS is an "open" trading system, meaning that allowance trading may take place with other sectors covered by the EU ETS, foreign trading systems, and "certified emission reductions" generated in developing countries under the Kyoto Protocol. If, for example, industrial emitters in Europe have less need for EU ETS allowances than they now project, those allowances could be sold in the secondary market, adding to supply and dampening prices.

The newspaper *The Australian*[50] cited an allowance price forecast of the EU open trading system by the Swiss investment bank UBS. UBS expected the EU price for allowances to average €5 per metric ton (approx. US$6.40) in 2012-2013. Further, UBS predicted that EU ETS allowances would remain in surplus until 2025. As of early January 2012, allowance futures prices for 2015 were settling at around €8.45, though they had ranged as high as €12.05 in September 2011. Some observers believe that businesses may have been acquiring and/or saving ("banking") emission allowances during the second EU trading period, 2008-2012, to use in the post-2013 period. Although rising allowance prices are generally expected, it remains very uncertain to what degree they may rise over the coming decade.

Alleged Effects on R&D and Investment in New Equipment

Airlines and A4A, the US industry trade group (formerly ATA), have raised concerns that the purchase of allowances may have an unintended consequence: limiting airlines' funds to invest in more fuel efficient aircraft and aircraft modifications, such as next-generation avionics, that could reduce emissions. Airlines and A4A have expressed specific concern that European government receipts from the EU ETS may not fund environmental projects or to help reduce emissions from aviation sources. They argue that aviation emissions reductions could be achieved more readily if the aviation industry could instead invest directly in new aircraft, retrofitting of existing aircraft, and alternative fuels.[51]

Potential for Profit Opportunities and Windfalls

The EU ETS could also provide carriers with a profit opportunity. Over the next nine years, aircraft operators will receive free allowances equal to 82%-85% of their historic CO_2 emissions of flights to and from EU airports. In total, the EC asserted that the free allowances could be worth €20 billion (US$27 billion) over the next nine years.[52] These allowances are assets that the carriers are free to sell, potentially using the proceeds to upgrade their fleets, acquire more fuel-efficient equipment or switch to lower-carbon biofuels which, in turn, would lower emissions per ton-kilometer, thereby reducing the need for allowances and reducing exposure to fuel expense volatility.[53]

Some analysts believe that carriers could make "windfall" profits from the freely distributed allowances under less competitive conditions. A recent study[54] by MIT researchers and others estimated that the cost impact of the EU ETS on US carriers would be small and that aviation operations would continue to grow. Airline profits could increase, especially if carriers are able to pass on all additional costs to their customers. This possibility, however, would depend on the degree of competition and the portion of the costs of allowances that airlines can pass through to customers. Windfalls could result if carriers increase ticket prices to reflect the prices of all allowances but only need to pay for a fraction of them, or are able to sell unneeded allowances. On the other hand, researchers concluded that profits would decline to the degree that carriers must purchase a greater share of their allowances and that competition increases.

It is too early to assess the degree to which carriers will be able to pass allowance prices through to customers, or to reduce their emissions growth. In January 2012, immediately following the inclusion of aviation sources in the EU ETS, some major US airlines began to impose $3 surcharges on passenger tickets for European flights, while German carrier Lufthansa indicates that it would increase its fuel surcharge to reflect the cost of purchasing carbon allowances later in the year.[55] China has ordered its airlines not to raise ticket prices to cover potential allowance costs.

Competitive Dynamics

Including aviation within the EU ETS may also affect competitive dynamics within the airline industry globally, particularly if the price of allowances were to rise significantly. In the air cargo sector, where leasing and subleasing of aircraft already are common, high allowance prices could encourage carriers subject to EU ETS to sublease space from contract operators that have sufficiently few EU flights that they are not required to obtain allowances. The same applies to chartered passenger flights. As small operators grow, however, they could exceed the *de minimis* criteria and become subject to allowance requirements.

Among airlines providing scheduled international passenger service, EU-based carriers will be required to obtain allowances for a far higher proportion of their total flights than non-European carriers. The non-EU carriers will be able to spread the cost of the allowances across a large volume of flights not subject to allowances, potentially gaining a competitive advantage.

If non-EU carriers were exempted from the EU ETS, as other countries have demanded, they could gain a further cost advantage over their European competitors. Arguably, this could distort competition in the EU market, and reduce the ability of European airlines to pass any additional costs through to customers.

The need to purchase allowances also has the potential to affect the entry of new carriers on routes between the United States and Europe. Because allowance allocations are based on historic emissions, they could disfavor carriers that wish to enter the EU market or to expand more rapidly than their competitors. (China, for example, argues that its carriers will be particularly disadvantaged for this reason.) To alleviate this potential problem, beginning in 2013, the EU will reserve 3% of annual aviation allowances for new or rapidly expanding operators.

Incremental Price per Trans-Atlantic Ticket

One benchmark for assessing the cost to aviation in the EU ETS is the amount it may add to the price of a trans-Atlantic airline ticket. The EU anticipates that the impacts of its ETS will be negligible. According to the European Commission,

> The inclusion of aviation in the EU ETS is expected to have a limited impact on ticket prices. Based on current carbon prices, the actual costs per ticket of a transatlantic flight would increase by less than 2 Euro [US$2.65], assuming that the value of the free allowances would not be passed to the passenger. If it were to be passed through, the ticket price could increase by around 12 Euro [US$15.85].[56]

A rough calculation by CRS supports the EU's estimate. According to the EU Commission, the per-passenger emissions on a flight between New York and London are estimated at 385 kg. At a hypothetical EU allowance price of €10 per metric ton, if all allowances needed to be purchased, the incremental cost would be about €3.85 per passenger (about $5 at current exchange rates, or over US$6 at the weakest dollar exchange rates of the past few years). However, as aircraft operators receive free 85% of allowances for 2012 and 82% for 2013-2020, the per-passenger costs in the near term could be less than $1 to $2 per passenger from New York to London, and proportionally more on longer flights. On most flights between the EU and the United States, the additional per-passenger cost will likely be less than $5 in the near term, given the high proportion of free allowances provided.

In the near term, the incremental costs of complying with the EU ETS are likely to be small per passenger. This is reflected in the initial responses by several airlines to their new inclusion in the scheme: By the beginning of January 2012, Delta Airlines began to test its ability to pass costs on to its customers by adding a $3 surcharge to transatlantic tickets, though Delta declined to say that this surcharge is linked to compliance with the EU ETS.[57] Soon after, additional airlines, including United and American, followed with identical surcharges, as have a variety of foreign airlines. The small size of the surcharges relative to other expenses and charges (e.g., baggage fees and fuel expenses) means that its effects are likely to be difficult to track as compliance and competition play out.

While the per passenger cost may be estimated to be relatively small, it could fluctuate in the future depending on the pricing of emission allowances. Current projections indicate that the market may remain relatively stable in the near term. Some airlines have voiced concerns about potentially volatile allowance prices, coupled with recent volatility in jet fuel prices. However, fuel price increases and allowance prices are likely to move in opposite directions,[58] with one tending to moderate the effect of the other.

The scale of the impact of the EU ETS may be illustrated by comparing the approximate costs of allowances and jet fuel: According to the Energy Information Administration, one barrel of jet fuel leads to emissions of about

0.40 metric tons of CO_2. In early 2012, the price of an allowance was 9 Euros, or about $12 per barrel of jet fuel. In comparison, the price of a barrel of jet fuel was almost US$140 per barrel, having risen from about US$50 per barrel in 2009.[59] By comparison, rising fuel expenses are a much greater expense for the aviation industry. The incremental cost of complying with the EU ETS in the near term may be around 5%-10% of the historic increase in fuel expense per passenger witnessed from 2002 to 2010 (*Figure 4).* Rising fuel expenses also require a rising share of airlines' revenues, from around 12%-15% in 2002 to around 26%-35% in 2010. Those higher fuel expenses alone increase incentives for airlines to improve efficiency and deploy biofuels that would, in turn, facilitate compliance with the EU ETS. On the other hand, to the degree that the cost of EU ETS compliance may be small, compared to rising fuel prices, the extra incentive of the EU ETS to reduce emissions would be comparably small.

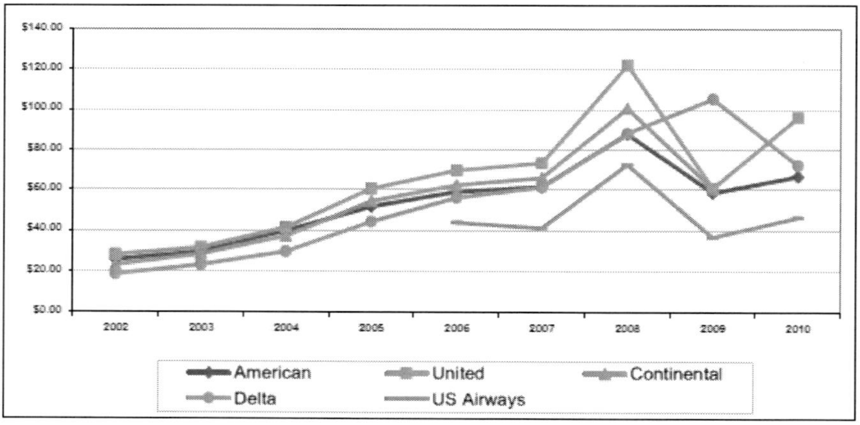

Source: CRS figure using data from AirlineFinancials.com, extracted January 24, 2012, http://www.airlinefinancials.com/airline_data_comparisons.html.

Notes: Data cover all flights, not just trans-Atlantic or long-haul flights. To the degree that trans-Atlantic flights use more fuel than the average, represented in this figure, their per-passenger fuel expenses would be greater than shown here.

Figure 4. Fuel Expense per Passenger 2002-2010 for Several USAirlines.

Significance of Extra EU ETS Costs to US Carriers

Even if the compliance with the EU ETS were to cost several dollars per trans-Atlantic ticket (a small percent change), compliance costs may be understood in the context of broader conditions in the aviation industry: strenuous competition, other rising costs, and strongly fluctuating profit margins.

First, airlines are concerned about effects that any higher ticket prices may have on demand for air services. Other experts estimate that any such impacts on air travel are likely to be small. Though passenger demand is sensitive to ticket prices, it is generally assessed to be somewhat "price inelastic" especially in long-haul aviation.[60] In other words, an increase in price would result in a proportionately smaller decline in customer demand. Sensitivity to prices varies strongly across fare classes, with first and business class travelers being less price sensitive than economy class travelers. Airlines may choose how to allocate allowance costs among fare classes, including allocating more of a price increase to less sensitive customers.

Second, the incremental costs could reduce operating margins. Faced with demand suppressed by continued economic weakness, coupled with market competition, companies may not be able to pass on the incremental costs of the EU ETS to customers. If they cannot pass on their compliance costs, they may experience lower operating margins and lower profits. If they can pass on the market value of all allowances, including the ones they receive free, their profits may expand. At expected allowance prices, this does not appear to be among the main factors to affect airline profits.

EXTERNAL OBJECTIONS TO COVERING FOREIGN-BASED AIRLINES IN THE EU ETS

The United States and several other non-European countries have voiced opposition to inclusion of their airlines in the EU ETS. Some countries say that EU regulation of their airlines violates their national sovereignty.

A principal argument is that the scheme requires allowances for the portions of flights to and from the EU that occur outside of EU airspace. This would include portions of flights over countries that are neither the origin nor destination. Several countries contend that this violates their sovereignty. Some say that the required allowances equate to a charge for using a third country's airspace—for example, when a flight from San Francisco to Paris flies over Canada.

Another effect is that a flight from San Francisco to Paris will require more emissions allowances than a flight from New York to Paris, even though the two flights may fly identical flight paths once they enter European airspace. Some allege that this could prompt airlines to stop over flights in countries near the EU (e.g., Morocco) rather than fly long-distance flights directly to or from EU airports.[61]

Non-European countries assert that the EU's unilateral action to include aviation provisions in the EU ETS undermines ICAO's role as the central forum to address GHG emissions from international aviation, as mandated in the Kyoto Protocol.

The EU has defended its position. EU officials assert that the EU ETS is consistent with the Kyoto Protocol, and that the Kyoto Protocol does not confer an exclusive mandate upon ICAO. They also state that bringing international aviation under the EU ETS does not contradict or undermine ICAO's role and ICAO's actions with regard to the mitigation of CO_2 emissions from aviation sources. In addition, EU officials contend that ICAO's 2004 decision ruling out the option of a *global* emission trading scheme for aviation left the EU free to impose its own emission trading scheme without violating ICAO agreements.[62] Further, the EU denies that the scheme infringes on national sovereignty, saying that the reach of the aviation provisions is analogous to US security requirements for departing flights that wish to land in the United States.

LITIGATION: AIR TRANSPORT ASSOCIATION OF AMERICA V. SECRETARY OF STATE FOR ENERGY AND CLIMATE CHANGE[63]

The Air Transport Association of America (ATA), along with several airlines based in the United States, challenged Directive 2008/101 in European courts, claiming that the EU law is in breach of both customary international law and various international agreements.[64]

The lawsuit was originally brought on December 16, 2009, before the High Court of Justice of England and Wales in the United Kingdom, the administering member State for the airlines involved in the lawsuit.[65] Because the claims challenged the validity of an EU law, the English court referred the case to the European Court of Justice for a preliminary ruling on July 8, 2010.[66]

The ATA[67] challenged the law on several grounds, claiming that the EU law violated the Chicago Convention, the Open Skies Agreement[68] between the United States and the EU, the Kyoto Protocol, and various principles of customary international law regarding airspace sovereignty and the freedom to fly over the high seas.[69] On December 21, 2011, the European Court of Justice, the highest court in the European Union, rejected these claims and upheld the law.[70] This ruling cannot be appealed and will be returned to the UK court to implement.

The ATA's Arguments

The ATA attacked the Directive on multiple grounds. It objected to the EU's unilateral imposition of a cap-and-trade system on airlines that are not based in the EU. The ATA argued that any environmental standards or market-based mechanisms to reduce GHG from aviation should be implemented through ICAO. The ATA contended that unilateral imposition could lead to the double regulation of aircraft emissions if other countries established their own measures for reducing aviation emissions. The ATA further objected to the Directive because it covers emissions that occur outside of EU airspace, when an aircraft is traveling to or from the EU on an international flight. Finally, the ATA claimed that the Directive's requirement to surrender allowances amounts to an impermissible tax or charge on airline fuel.

Summarizing the ATA's arguments, the Advocate General for the European Court of Justice stated:

> First, they contend that the European Union is exceeding its powers under international law by not confining its emissions trading scheme to wholly intra-European flights and by including within it those sections of international flights that take place over the high seas or over the territory of third countries. Secondly, they maintain that an emissions trading scheme for international aviation activities should be negotiated and adopted under the auspices of the ICAO; it should not be introduced unilaterally. Thirdly, they are of the opinion that the emissions trading scheme amounts to a tax or charge prohibited by international agreements.[71]

Opinion of the European Court of Justice

The European Court of Justice decided this case by answering two questions. First, the court examined whether the plaintiff airline companies and associations, as individuals rather than state actors, could use any of the international agreements or principles of customary international law to challenge the Directive's validity. Second, the court examined whether international agreements and customary international law would invalidate the Directive if the ATA had rights to challenge it.

In answering the first question, the court found that the ATA could not challenge the Directive under the Chicago Convention, because the EU is not a Party to the Convention, or the Kyoto Protocol, because the Kyoto Protocol does not grant an individual an interest that could be raised before the court. The court found, however, that the ATA could challenge the Directive under parts of the Open Skies Agreement and customary international law. In answering the second question, the court determined that the Directive is valid under the Open Skies Agreement and the principles of customary international law.

Chicago Convention and Kyoto Protocol: The ATA Cannot Use the Chicago Convention or Kyoto Protocol to Challenge the Validity of Directive 2008/101

The court reasoned that for an individual to be able to challenge the validity of an EU law on the grounds that it violates a separate international agreement, the EU must be bound by the international agreement.[72] The court determined that although all 27 of the European Union member States are bound by the Chicago Convention, the EU, as a distinct entity, has not signed the agreement.

The EU merely holds an observer status under the Chicago Convention. The court determined that the EU is therefore not bound by the Chicago Convention and the ATA could not use the Chicago Convention to challenge the validity of an EU law.[73]

The court found that the EU is bound by the Kyoto Protocol. However, the court reasoned that in order for an individual to use an international agreement to challenge the validity of an EU act, the "nature and broad logic" of the international agreement "must be capable of conferring rights which an individual can invoke before the courts."[74] Furthermore, the specific provisions relied upon must be "unconditional and sufficiently precise to enable an individual to invoke them before the courts."[75] On this basis, the court rejected ATA's arguments raised under the Kyoto Protocol. The ATA argued that, under Article 2(2) of the Kyoto Protocol, the EU must only seek to reduce emissions from aviation activities by working through the ICAO. However, the court noted:

> [T]hat provision, as regards its content, cannot in any event be considered to be unconditional and sufficiently precise so as to confer on individuals the right to rely on it in legal proceedings in order to contest the validity of Directive 2008/101. Consequently, the Kyoto Protocol cannot be relied upon in the context of the present reference for a preliminary ruling for the purpose of assessing the validity of Directive 2008/101.[76]

Thus, the ATA could not rely upon the Chicago Convention or the Kyoto Protocol to challenge the validity of the EU law.

Ruling: Directive 2008/101 Does Not Violate the Open Skies Agreement or the Principles of Customary International Law

Despite disallowing the ATA to use the Chicago Convention or the Kyoto Protocol to challenge the validity of the Directive, the court determined that the ATA could use Article 7 (aircrafts shall be subject to the laws of a state while it is within that state's territory), portions of Article 11 (customs taxes or fees cannot be imposed on fuel that is onboard an aircraft), and a portion of Article 15(3) (environmental regulations must be nondiscriminatory and compliant with ICAO standards) of the Open Skies Agreement to challenge the Directive's validity.[77] Furthermore, the ATA could challenge the Directive pursuant to the customary international law principles of (1) complete state sovereignty over its airspace; (2) freedom of flight over the high seas; and (3) the high seas are free from state sovereignty.[78] However, the court found that Directive 2008/101 is consistent with the Open Skies Agreement and customary international law.

The court disagreed with ATA's claim that Directive 2008/101 violates Article 7 of the Open Skies Agreement and various principles of customary international law because it regulates aircraft both when they are entering and departing the European Union, and when they are flying over the high seas or a third country's territory.[79] The court declared that the Directive does not impermissibly regulate aircraft extraterritorially. The court noted that the trading scheme is not triggered by aircraft operators flying over the high seas or over the territory of any country. Rather:

> [i]t is only if the operators of such aircraft choose to operate commercial air routes arriving at or departing from aerodromes situated in the territory of the Member States that, because their aircraft use such aerodromes, those operators are subject to the allowance trading scheme... It follows that Article 7(1) of the Open Skies Agreement does not preclude the application of the allowance trading scheme set up by Directive 2003/87 to operators of aircraft, such as airlines established in the United States, when their aircraft engage in flights which arrive at or depart from [the European Union].[80]

The court's reasoning also led it to dismiss ATA's arguments based on customary international law. It reiterated that the mere fact that the number of allowances is determined by the full length of a flight does not mean that the Directive regulates aircraft operators outside the EU's jurisdiction.[81] Again, the court emphasized that the Directive only regulates aircraft that choose to land at or depart from an EU aerodrome and does not regulate aircraft that merely fly over the high seas or through another country's airspace. The court determined the Directive to be valid under these principles of customary international law.[82]

With regard to Article 11 of the Open Skies Agreement, the court disagreed with ATA's position that the Directive, which calculates how many allowances must be surrendered by measuring how much fuel has been burned, imposes an impermissible tax or duty levied on airline fuel.[83] The court declared that the law "constitutes a market-based measure and not a duty, tax, fee or charge on the fuel load."[84] The court distinguished the trading scheme from a tax by noting that an airline, under the trading scheme, could potentially make a profit by burning less fuel and selling excess allowances.

The court also found the Directive to be compliant with Article 15(3) of the Open Skies Agreement. The ATA argued that the Directive is incompatible with ICAO environmental standards and impermissibly limits the volume and frequency of air traffic in contravention to Article 15(3).[85] However, the court again dismissed ATA's arguments and ruled that the Directive does not infringe on any ICAO environmental standards, finding that there is no indication that the "implementation of market-based measures ... would be contrary to the aviation environmental standards adopted by the ICAO."[86] The ATA further argued that, because the EU acted unilaterally, the Directive could lead to double regulation if another country imposed its own market-based mechanisms on the aviation sector. Although an ICAO resolution indicates that market-based mechanisms "should not be duplicative," the court noted that the Directive contains provisions that enable the EU to alter its market-based system to ensure that airlines are not doubly regulated when operating on international routes.[87]

The court noted that under Article 15(3) of the Open Skies Agreement, a Party is permitted to limit the volume and frequency of air traffic for environmental reasons, but such environmental measures must be nondiscriminatory.[88] Because the Directive regulates both international and domestic commercial flights that arrive in or depart from the EU in the same manner, the court held that the Directive is not discriminatory and "not invalid in light of Article 15(3) of the Open Skies Agreement."[89]

Because the Court of Justice has upheld the validity of Directive 2008/101, the law will remain in force and airlines using EU airports will be required to participate in the EU ETS.

US Government Actions

The United States government has objected to the inclusion of aviation in the EU ETS and warned that inclusion of US airlines in the scheme could spark protracted trade disputes. A December 16, 2011, letter from Secretary of State Hilary Clinton and Transportation Secretary Ray LaHood to European Commission officials asserted that the EU ETS is inconsistent with international law and urged the EU to work with the broader international community in the context of the ICAO forum to address the challenges of reducing aviation emissions.[90]

Also on December 16, 2011, the US Department of Transportation (DOT) issued an order requiring nine major European airlines and seven US airlines to submit detailed data on the free allowances allocated under the EU ETS, as well as passenger and cargo data.[91] The US carriers are also required to submit information regarding any amounts paid to administering European countries under the EU ETS for the purchase of auctioned carbon allowances. A4A expressed its approval of the action. A4A noted that the information can be used to assist in its arguments against the application of the EU ETS to US airlines, although it remains unclear how the data could specifically be used.[92]

As is typical in national environmental laws, the EU ETS law provides for penalties against aircraft operators that do not comply with its requirements. For its part, the United States has responded by threatening trade sanctions if the EU makes an attempt to force foreign airlines to comply with the emissions trading system.[93]

Conceivably, the Administration could agree to an ICAO instrument that could be achieved under existing US authorities and appropriated resources. For example, the US Environmental Protection Agency (EPA) has authority under the Clean Air Act to regulate emissions, including GHG emissions, from aircraft. It already uses this authority to bring US aircraft emission standards (for example, for nitrogen oxides) in line with those agreed under ICAO. It could also do so for CO_2. The Administration has shown no indication, however, that it would agree to any requirements that are not: "technology-following";[94] agreed with FAA, EPA, and other relevant agencies (e.g., the Navy on biofuels); and well accepted by the US aviation industry, as in the past.

The Article 84 Option

A possible course of action for countries opposing the EU ETS is to file a formal complaint before ICAO in what is referred to as an Article 84 filing. Article 84 of the Chicago Convention establishes a formal ICAO process for settling differences among member countries. Under ICAO rules, the complaining state would submit an Article 84 application including detailed facts, supporting data and legal arguments, and relief sought. The member country against which the complaint is levied would be informed of the complaint and provided with a copy of the complainant's submittal and offered the opportunity to present a counterclaim. Under a timetable set by the ICAO Council, the parties would be given the opportunity to file additional pleadings and make oral arguments, which may include the testimony of witnesses or experts. Ultimately, the Council, after hearing oral arguments in the case, would render a decision. The proceedings may be discontinued if the applicant provides written notification that it does not wish to proceed, unless the respondent specifically objects within a specified timeframe set by the Council. As previously noted, the EU is not a member of ICAO. Therefore, any Article 84 filing in the case of the EU ETS would have to be levied against one or all of its member States.

International Actions

Many countries have opposed inclusion of international aviation in the EU ETS. On July 28, 2011, member countries of the Latin American Civil

Aviation Commission signed a joint declaration urging the EU not to unilaterally impose its ETS on foreign flights without prior agreement with affected countries. The countries requested that ICAO continue its work toward a framework for market-based criteria for aircraft emissions reductions or greenhouse gas limitations.

Subsequently, on September 30, 2011, 26 ICAO member countries including the United States, China, India, Japan, Korea, Russia, Mexico, and several Latin American countries, signed a joint "Delhi Declaration" at the New Delhi meeting of the ICAO Council opposing inclusion of international aviation in the EU ETS. Instead, the declaration supports ICAO efforts to develop meaningful aircraft carbon emissions standards with a target of adopting such standards at the General Assembly in 2013. It urges the EU to refrain from including non-EU flights in the EU ETS and to work collaboratively with the international community. Besides EU Member States, a number of countries did not join the declaration: Australia, Canada, New Zealand, and others.

Since the aviation component of the EU ETS began in January 2012, some governments have instructed their airlines not to participate, threatened to sue the EU, or suggested they might impose trade sanctions. Less publicly, there are indications that some governments may be negotiating to explore exemptions for their airlines based on "equivalent measures" or other arrangements.[95] For example, the Chinese government unveiled a plan to reduce its airlines' CO_2 emissions intensity (emissions per revenue metric ton-kilometer) to 22% below 2005 levels by the end of 2020.[96] It may be exploring this policy as a basis for an equivalent measures exemption.

Indian sources have suggested that they may follow China's lead in instructing their airlines not to comply or in threatening retaliatory fees or other reactions.[97] Though the Indian government in 2011 asked carriers not to submit their emissions data to the EU, the EU reports, however, that they remain in compliance with reporting requirements.

Representatives of 26 countries, including China, India, Japan, Russia, the United Arab Emirates, and the United States, are meeting to seek a common response to the EU inclusion of international aviation. For example, in February 2012, many of those countries identified a variety of possible responses, including prohibiting non-EU airlines from participating in the EU ETS, refusing to discuss operating rights with EU aircraft operators, imposing retaliatory taxes on EU airlines, or exercising Title 84 Settlement of Differences rules under ICAO (see "The Article 84 Option").[98] Those governments said, however, that they would not undertake a joint response;

each government would select its own appropriate responses. According to news reports, the Department of State has said it would be "premature" to discuss specific retaliatory measures.[99]

An EU official noted that "any measures that discriminate against European companies will obviously be taken to ... international court."[100] Senior EU officials have indicated that a global treaty under ICAO that includes emission targets and "robust measures" could lead to EU flexibility in applying the EU ETS law. They see the ICAO aim to propose global market-based measures by the end of 2012 as an avenue for resolving the conflict.

ACTIVITIES IN ICAO

ICAO recognizes the Kyoto Protocol requirement for industrialized nations to reduce greenhouse gas emissions. Also, the 1997 Kyoto Protocol (to which the United States is not a Party) specified that the industrialized country Parties should pursue limitation or reduction of greenhouse gas (GHG) emissions from aviation bunker fuels, working through ICAO.[101]

Through the 2000s, the EU has warned that, without stronger action under ICAO, it would include the aviation sector in the EU ETS. In October 2011, faced with the 2012 implementation of the EU's cap on aviation emissions, the ICAO Council agreed to accelerate its work and continue to explore market-based measures to "reach a global solution." *Appendix B* describes the major related activities and policies of ICAO to address GHG.

Latest ICAO Policy

In November 2011, the ICAO Assembly agreed to an aspirational goal of improving fuel efficiency by 2% annually and stabilizing its global CO_2 emissions at the 2020 levels. The ICAO Assembly has committed to making further progress, focusing on (1) States' action plans and assistance to States, (2) sustainable alternative fuels for aviation, (3) exploration of global schemes for market-based measures, and (4) global aspirational goals, in order to moveinternational aviation closer to a sustainable future. It says that it has shifted from a "standards and policies setting" phase to an "implementation" mode, based on voluntary submission of action plans by member States.

Also in November 2011, a majority of ICAO member States adopted a declaration urging the EU and its member States to refrain from including international aviation operations to or from EU member States in the EU emissions trading system. The EU member States, Australia, New Zealand, and several other countries submitted reservations against this resolution or did not vote to support it. By consensus, the member States confirmed the important role of ICAO and urged acceleration of ICAO's work on market-based measures (MBMs). The ICAO statement to the 2011 UNFCCC negotiations in Durban urged:

> It is of utmost importance that market-based measures for international aviation should be treated as one element of ICAO's comprehensive mitigation strategy to achieve the global aspirational goals, as part of global solutions for the sustainable future of international aviation, and not in isolation.[102]

To this end, ICAO agreed to develop by the end of 2012 "a clear roadmap towards the development of global solutions."[103]

CONGRESSIONAL ACTION

Despite objections by the US and other governments, the EU has stated that flights to and from the EU will remain covered by the EU ETS unless they are exempted due to "equivalent measures" in other countries, or unless the EU determines that a global, binding agreement has been reached under ICAO or the UNFCCC that would reduce aviation emissions by at least as much as the EU ETS. Aviation representatives have raised the issue to Members of Congress.

Members of Congress have objected to coverage of US international aviation by the EU ETS and may consider further legislative responses to the EU action. A sense of the Congress resolution was agreed by House and Senate conferees and was included in the FAA Modernization and Reform Act of 2012 (P.L. 112-95). Specifically, Section 509 of the act expresses the sense of the Congress that implementation of international aviation provisions of the EU ETS is inconsistent with the Chicago Convention on International Civil Aviation and other relevant aviation agreements, and is antithetical to building international cooperation to address effectively GHG emissions from international civil aviation. It expresses a view that the EU should instead work through ICAO to develop a consensual approach, and holds that the US government should use all available political, diplomatic, and legal tools to ensure that the EU ETS is not applied to US-registered aircraft.

Other bills in the House and Senate seek more specific actions to exclude US aircraft from participating in the EU ETS. In October 2011, the House of Representatives passed H.R. 2594, the European Union Emissions Trading Scheme Prohibition Act of 2011. It would prohibit US aircraft operators from participating in the EU ETS. It also would direct the Administration to negotiate and take other actions to ensure that US civil aircraft operators are "held harmless" by any unilaterally imposed EU regulation.

A companion Senate bill, S. 1956, was introduced but saw no action in 2011. The Senate bill is similar to H.R. 2594. However, it would give greater flexibility to the Secretary of Transportation to prohibit US carriers from complying with the EU ETS, basing it on a determination that doing so would be in the public interest.

Among the concerns expressed in both bills are that:

- The EU scheme was imposed unilaterally.
- Non-EU aircraft operators must have allowances to cover their emissions over non-EU countries and in international airspace for flights to and from the EU.
- The EU's coverage of international aviation emissions undermines cooperation under ICAO.
- The EU law does not require administering States to use the revenues collected from the aviation allowances for environmental purposes in the same sector.
- EU responses to the objections raised by the US government have been deemed unsatisfactory.

If either of the bills were to be enacted, the language could pose challenges for US aircraft operators serving the EU. Were the Secretary of Transportation to prohibit them from participating in the EU ETS, that could be construed as an instruction not to comply with the laws of the EU and its member States. This could prevent those operators from receiving their shares of allowances.[104] It also could subject those operators to non-compliance penalties, including exclusion from the EU aviation market.

Several courses of action for Congress seem possible, perhaps in conjunction with the Administration:

- prohibit US aircraft operators from participating in the EU ETS and let play out any non-compliance actions on the part of EU administering authorities and US Administration responses;
- continue political pressure on the EU and its member States to amend their laws governing inclusion of aviation in the EU ETS;
- pass a resolution encouraging ICAO to accelerate work on adoption of a global trading scheme, CO_2 limits on aircraft engines, or other actions to limit international aviation emissions;
- encourage the Administration to pursue action under the World Trade Organization(WTO), although the basis for action may be undetermined;[105]
- support domestic programs, under existing or new authorities, that the EU might accept as "equivalent measures" and facilitate economical compliance of US aircraft operators with the EU ETS requirements, such as continued development and deployment of biomass-based aviation fuels, or hypothetically, a US emissions charge-and-rebate scheme;
- support domestic actions that provide benefits to the United States, such as improvements in air traffic management, research in more fuel efficient equipment design, etc., that would also reduce US aviation emissions. Such actions might be bundled into the voluntary national program encouraged by ICAO and negotiated with EU as justification to exempt US carriers arriving in EU airports.

Other options may be proposed as the debate evolves.

CONCLUSION

1. The European Union is bound by international and domestic laws to reduce its greenhouse gas emissions to 20% below 1990 levels by 2020. As of 2009, aviation was the largest source of GHG emissions not yet covered by measures designed to reduce GHG emissions. As the EU did not foresee effective global measures emerging from cooperation under ICAO, as mandated under the 1997 Kyoto Protocol, the EU extended its Emissions Trading Scheme (EU ETS) to control aviation emissions effective January 1, 2012. (Australia and New Zealand have also included aviation in their domestic programs, but not international flights to and from their countries.)

2. The European Court of Justice (ECJ) has ruled that including aviation in the EU ETS is lawful according to international customary law and bilateral aviation treaties. As the ECJ is the highest court in the EU, there is no appeal to this decision. As such, the EU has proceeded to implement its law, as have EU member States, which have transposed EU law into their national laws.
3. Reportedly, all major US aircraft operators have been complying with the EU ETS requirements (reporting requirements began in 2010). None has said publicly that it will not continue to comply in the system. A checkpoint will occur on April 30, 2013, when aircraft operators must submit allowances sufficient to cover their emissions in 2012 for flights to and from the EU ETS countries.
4. H.R. 2594 (passed by the House in October 2011) and S. 1956, the European Union Emissions Trading Scheme Prohibition Act of 2011, if enacted, would direct the Secretary of Transportation to prohibit US aircraft operators from complying with the EU ETS. This could force them into making a choice of breaking either US or EU law and subject them to financial or even access penalties. Congress may further consider and take legislative action in 2012.
5. The United States Government has opposed unilateral imposition of the EU ETS on emissions from flights, or portions of flights, that occur outside of EU airspace. China, India, Russia, and other countries, as well as many aviation associations, have also objected strenuously to coverage of international aviation. China has threatened that its carriers will not pay for emission allowances if they are needed to comply with the system
6. As the International Air Carriers Association (IACA) has stated, "... the real issue is political, not legal."[106] The airline industry foresees resolution ultimately found working through ICAO, not the courts. The EU continues to insist that it would prefer expeditious agreement under ICAO to an effective *global* scheme to reduce international aviation emissions, but that it will implement its laws in the absence of a multi-lateral agreement.
7. ICAO has agreed to accelerate some of its work on CO_2 standards for new aircraft, "drop-in" biomass fuels for aircraft, voluntary national programs to address emissions from aviation, other research into more efficient aviation technologies, and market-based measures (MBMs)—potentially including emissions trading systems. It remains unclear whether any ICAO agreements would be "aspirational" and

voluntary, as past decisions, or whether they would have some degree of legal force to assure implementation by all Members. The EU insists that any ICAO must be a treaty, having binding requirements.

8. Additional questions about the future work of ICAO arise: first, whether ICAO will maintain its distinction of responsibilities between "developed" and "developing" countries—a bifurcation that the Durban Platform[107] under the UNFCCC has apparently removed; and, second, how ICAO may pursue apparent agreements in its resolutions that financing would be provided to developing countries for their actions to abate aviation emissions.
9. The EU intends to continue to implement the aviation components of the EU ETS, although it has left open the possibility of exempting certain flights. In particular, the EU may exempt flights arriving in EU airports from other countries that adopt "equivalent measures" to address aviation emissions as long as those measures are non-discriminatory.
10. One EU official stated that the EU would agree to suspend inclusion of aviation in the ETS only if a new global ICAO scheme met three conditions: it must deliver more emissions reductions than the EU ETS on its own; it must have targets and measures; and any action must be non-discriminatory and apply to all airlines. Alternatively, EU law allows exemptions for incoming flights from countries that have adopted "equivalent measures" to reduce emissions.
11. The political conflict on aviation emissions could decline over the next two years as aircraft operators improve their understanding and ability to comply with the EU ETS requirements. Various trends are likely to influence positions, including airlines' efforts to improve fuel efficiency, and growth of demand for air services if or as economic recovery continues. Current conflict also may be resolved by multilateral agreements under ICAO or the United Nations Framework Convention on Climate Change. Alternatively, the conflict could continue, with potential financial sanctions, retaliatory trade actions, and economic ramifications.
12. Additional strategies to reduce or resolve the dilemma may exist. For example, it is unclear whether options have been explored bilaterally between the United States and the EU regarding types of measures that might provide domestic benefits, in the United States' view, and satisfy the EU's goal of reducing aviation emissions. New options might facilitate resolution of the conflict between the two

governments. For example, some options in the United States may exist that would yield domestic benefits, notably improving air traffic efficiency, or supporting new infrastructure for aviation biofuels, that would also reduce aviation emissions.

APPENDIX A. NOTABLE DATES FOR AVIATION IN THE EU ETS

19 November 2008	EU Directive 2008/101/EC amends the EU ETS Directive (2003/87/EC) to include aviation activities
9 February 2009	The EU Commission publishes its preliminary list of aircraft operators included in the EU ETS
2 August 2009	The Commission publishes its Guidance on Annex I activities and its Decision on historic aviation emissions to be used for benchmarking aviation emissions in the EU ETS
1 January 2010	Start of first monitoring period of aviation activities and emissions, for determining historical annual emissions by aircraft operators
1 January 2011	Start of second monitoring period for determining annual emissions
31 March 2011	Deadline for aircraft operators to submit 2010 emissions data and 2010 benchmark data
1 April 2011	EU ETS expands to include Iceland, Lichtenstein, and Norway (EFTA, or known collectively with the EU as EU-EFTA)
20 April 2011	EU publishes updated list of aircraft operators subject to the EU ETS (Reglement394/2011)
26 September 2011	EU Commission sets the ton-km benchmark used to allocate emissions allowances to aircraft operators
1 January 2012	Start of third monitoring period for annual emissions; beginning of first compliance period with emissions limits
28 February 2012	Administering Authorities in EU Member States issue allowances for 2012 to those aircraft operators that successfully applied for them
11 June-29 June 2012	ICAO 196th Council Phase

17 September-5 October	ICAO Council Session—Committee Phase 2012
29 October-16 November	ICCAO 197th Council Phase2012
31 March 2013	Aircraft operators' 2012 Annual Emissions reports due
30 April 2013	Final date for covered aircraft operators to surrender allowances for 2012 emissions to Administering Authorities

APPENDIX B. ICAO AND ITS APPROACH TO ADDRESSING GHG EMISSIONS

Background on ICAO

Under the Chicago Convention, States agree to undertake "to collaborate in securing the highest practicable degree of uniformity in regulations ... in all matters in which such uniformity will facilitate and improve air navigation."[108] The Chicago Convention also prohibits States from relying upon the State of registration of aircraft as a basis for discriminating, either in promulgating laws affecting international civil aviation or in allowing access to airports.[109]

ICAO develops non-binding standards, guidance, and policies for States and the aviation industry to use. ICAO does not have regulatory authority. Once it adopts guidance or standards, it is up to member States to enact ICAO's recommendations into law or regulation and to enforce them, as appropriate. ICAO has dispute resolution procedures when disagreements arise on how States may be implementing their regulations or ICAO's resolutions.

ICAO's Approach to Addressing GHG Emissions

ICAO has focused on four approaches to address GHG emissions from international aviation:

- development of data and scientific background;
- market-based measures:
  - emissions levies (or fees);

    - emissions trading systems;
    - voluntary measures, including national action plans;
- technological solutions; and
- operational measures.

Most of this work is conducted by the 21- member Committee on Aviation Environmental Protection (CAEP),[110] which recommends standards that may be adopted by the ICAO Council (the executive committee[111]) and Assembly (comprising delegates of all 190 member States). The ICAO standards are not legally binding on the United States until they are promulgated as regulations under the Clean Air Act by the US Environmental Protection Agency. (They are enforceable by FAA.)

To assist with its mission to address GHG emissions, ICAO also established in 2007 the Group on Action on International Aviation and Climate Change (GIACC) to recommend an aggressive program of action to ICAO.

Data and Scientific Background

ICAO produces through expert groups a number of data sets and documents to underpin its consideration of guidance, standards, and recommendations.

To this end, ICAO holds conferences and workshops and issues documents that help build a common reference set among stakeholders.

Operational Measures

ICAO's CO_2 reduction strategy aims at improvement of Air Traffic Management (ATM). It has issued the Global Air Navigation Plan to assist States and regional planning organizations. It includes such emissions-related aspects as flexible use of airspace, air traffic flow management, terminal area design and management, performance-based navigation, and other options. ICAO has also produced guidance in its Operational Opportunities to Minimize Fuel Use and Reduce Emissions. CAEP is in the process of producing updated guidance.

Emissions Fees

On emission-related levies, ICAO made a policy statement in 1996 strongly recommending that any environmental levies be in the form of charges (i.e., on actual emissions performance) rather than taxes, and that the funds collected should be applied in the first instance to mitigating the environmental impact of aircraft engine emissions. Such charges should be based on the costs of mitigation to the extent that costs can be properly identified and directly attributed to air transport. In 2001, the Assembly recognized the continuing validity of this policy, urged States to refrain from unilateral action to introduce emission-related levies inconsistent with the policy, and called for further studies and the development of further ICAO guidance on the subject. In 2004, ICAO[112] reiterated its 1996 resolution on emission-related levies[113] on aviation and urged Member states to refrain from imposing them unilaterally until at least the following ICAO Assembly in 2007.

Emissions Trading

In 2004, the ICAO Assembly endorsed further development of an open[114] emissions trading scheme for international aviation, with further work focused on two alternative approaches: one for a voluntary trading system, and one to "provide guidance for use by Contracting States, as appropriate, to incorporate emissions from international aviation into Contracting States' emissions trading schemes consistent with the UNFCCC process."[115] In short, the ICAO resolution anticipated that member States might cover international emissions in their own emissions trading schemes. In 2007, a resolution ruled out the option of a global emission trading scheme for aviation. Instead, it produced guidance for member States should they include international aviation in their own emission trading schemes. ICAO urged that inclusion be with "mutual agreement" and non-discriminatory.

ICAO's efforts on MBMs include a methodology for calculating carbon emissions from aviation sources for use in carbon offset programs.

Voluntary Measures

Aspirational Global Goals

The ICAO Assembly agreed[116] that States will work through ICAO to achieve a global annual fuel efficiency improvement rate of 2% for the medium-term (up to 2020) and an aspirational global annual fuel efficiency improvement rate of 2% through 2050. The Member States agree that this goal would not be attributed to individual States and that each State may voluntarily contribute to reaching the goal.

In addition, ICAO resolved that it and its Member States would "work together to strive to achieve a collective medium-term global aspirational goal of keeping the global net carbon emissions from international aviation from

2020 at the same level," taking into account a variety of circumstances, from growth in the international aviation industry to the respective capabilities of Member States.[117]

National Action Plans

ICAO's program of action urges member countries to adopt action plans to address, in particular, CO_2 emissions from international civil aviation activities with the goal of establishing comprehensive plans based on the roadmap for action through 2050 adopted by the ICAO General Assembly in 2010. ICAO's program helps identify appropriate measures to reduce emissions from international aviation operations, monitor progress toward achieving worldwide goals. Under a 2010 ICAO agreement, member States are invited to voluntarily submit national action plans to reduce CO_2 emissions from aviation. Those plans may be submitted annually, with the first by the end of June 2012.

The voluntary action plans are expected to contain, at a minimum, information on (1) national contacts; (2) expected results of the plan, ideally to 2050; (3) a list of measures; and (4) information on any needs for assistance.

Financial Assistance to Low Income Countries

ICAO has agreed to help low income States with specific needs to address GHG emissions, through targeted technical and financial assistance. Because GHG emissions from aviation are not covered by the Kyoto Protocol, no financing is available through the Kyoto Protocol mechanisms.

ICAO weighed in on one option for international financing of measures by low income countries under the UNFCCC, as put forward by the UN Secretary General's High-level Advisory Group on Climate Change Financing (AGF).[118] That option would generate revenues with MBMs, such as carbon or passenger fees, applied to international aviation. In particular, ICAO stated that,

> It should be noted that the global aspirational goals for the international aviation sector, adopted by the 37th Session of the ICAO Assembly, will require adequate financial resources within the sector itself, enabling it to effectively respond to the global climate change challenge.

In 2009, the ICAO High Level Meeting stated, "further elaboration on measures to assist developing States and to facilitate access to financial resources, technology transfer and capacity building."[119] ICAO asserts that it is the appropriate institution to deal with aviation financing. It can adapt the financial instruments to aviation specific goals and assist developing countries. It can provide assistance not only financially, but also through technology transfer and capacity building.

Technology Solutions

Since the 1997 mandate in the Kyoto Protocol, ICAO has produced technical information and a variety of voluntary options and recommendations. In 2001, the ICAO Assembly[120] called for emphasis on technical solutions to address GHG emissions while continuing consideration of market-based measures. ICAO takes a "technology following"[121] approach to standards, embodying technologies that are widely in use (i.e., a majority of users) and encouraging their adoption by remaining manufacturers and operators.

ICAO has focused on two technological areas for reducing GHG emissions: engine certification standards and alternative aviation fuels. In 2010 ICAO adopted its most recent emission certification standards for nitrogen oxides (NO_x), which contribute to formation of ozone (a local air pollutant) and GHG. The US EPA is in the process of promulgating regulations in line with ICAO's standards.

ICAO also agreed in February 2010 on a plan to develop an aviation CO_2 emission "certification standard" for aircraft engines, for consideration at the end of 2013. The Secretary General has indicated that the organization would release a proposal at the end of 2012, an acceleration of the process. This standard would cover *reporting* of CO_2 emissions from engines; it is not expected to address *limitation* of CO_2 emissions. Before any standard would be binding in the United States, the EPA would need to promulgate a corresponding regulation.

Propelled by rising and volatile prices for fossil-based aviation fuels, ICAO and its member States have given increasing support to development and demonstration of biomass-based fuels. ICAO produced a Global Framework for Aviation Alternative Fuels, and endorses the use of sustainable alternative fuels for aviation, particularly use of "drop-in" fuels in the near- and medium-term.

ICAO has identified the need for financing of alternative fuel infrastructure and incentives. To that end, it is discussing a possible framework with the World Bank and the Inter-American Development Bank.

End Notes

[1] Greenhouse gases (GHG) include dozens of compounds that, through various processes, alter the incoming and outgoing energy of the Earth system. The changing energy balance induces climate change. The principal human-related GHG is carbon dioxide (CO_2), and about three-quarters of human-related CO_2 emissions result from combustion of fossil fuels. So, while scientific and policy concerns surround all GHG, the focus of this report and of the EU regulation of aviation GHG emissions is CO_2. However, several other types of emissions from aviation also affect the climate, including water vapor, nitrogen oxides, particulate matter, hydrocarbons, and others.

[2] An explanation of Signatories, Parties, entry into force, and a few additional terms of international treaties may be helpful: When nations negotiate an international agreement or treaty, the agreement is opened for signature. The Signatories often participate in further activities pursuant to the agreement until the agreement enters into force, typically after a majority or qualified majority of Signatories have deposited their instruments of ratification, acceptance, approval, or accession (depending on each nation's law) with the organizational host of the agreement (e.g., the United Nations). At the point that the agreement enters into force, only the nations that deposited such instruments become Parties to the treaty. Only these Parties are bound by the terms of the agreement and may participate (i.e., with speaking and voting rights) in further activities under the treaty. Non-Parties may be allowed to participate as Observers. For example, the United States is a Signatory to the Kyoto Protocol but declined to ratify it, and so is not a Party; the United States has official Observer status. As such, the United States is not bound by the Kyoto Protocol. Typically, treaties include provisions that allow a Party to withdraw from it. For example, Canada has announced its intention to withdraw from the Kyoto Protocol; at such time, it would no longer be a Party to it.

[3] Conference of the Parties to the United Nations Framework Convention on Climate Change. "The Cancun Agreements: Outcome of the work of the Ad Hoc Working Group on Long-term Cooperative Action under the Convention," Decision 1/CP.16. 2010.

[4] See, for example, America's Climate Choices: Panel on Limiting the Magnitude of Future Climate Change; National Research Council. Limiting the Magnitude of Future Climate Change. Washington, D.C.: The National Academies Press, 2010.

[5] For background on the international negotiations to address climate change, see CRS Report R40001, A US-Centric Chronology of the International Climate Change Negotiations, by Jane A. Leggett.

[6] Emissions from international maritime transport pose a similar challenge.

[7] Article 2.2 of the Kyoto Protocol.

[8] The Provisional International Civil Aviation Organization replaced the International Commission for Air Navigation (ICAN), which had been in existence since the early days of powered flight. When the Provisional International Civil Aviation Organization became the ICAO, it included additional member countries and was placed under the umbrella of the United Nations (UN). It presently resides organizationally within the UN as a specialized agency under the jurisdiction of the UN Economic and Social Council.

[9] Although aviation emits a number of compounds that influence local and global climates, only aviation's carbon dioxide (CO_2) emissions are covered by the EU ETS. For some sectors, the EU ETS limits other GHG as well.

[10] Information on the Australian control scheme is available from their Department of Climate Change and Energy Efficiency: http://www.climatechange.gov.au. New Zealand's policies

are available from the Ministry for the Environment: http://www.mfe.govt.nz/issues /climate/policies-initiatives/index.html#nzets.

[11] Delbeke, Jos. "A New Flightplan - Getting Global Aviation Climate Measures Off the Ground." Keynote Address presented at the conference, A New Flightplan - Getting Global Aviation Climate Measures Off the Ground, February 7, 2012. http://ec.europa.eu/clima/ news/docs/speech_en.pdf.

[12] Ibid.

[13] Airlines for America (A4A) was called the Air Transport Association (ATA) until mid-2011. Both names are used throughout this report, depending on the context and timing of events in the discussion.

[14] Technically, aviation emissions were about 5% of the human-related "radiative forcing," which is the influence on global average temperature at the top of the troposphere. Lee et al., op. cit. This estimate was produced by the scientists who authored the aviation section of the 2007 assessment report of the Intergovernmental Panel on Climate Change, on which ICAO also relied for its published analyses. The 2007 IPCC report concluded that the radiative forcing of aviation in 2005, but based on 2000 operations data, was 14% lower.

[15] In this instance, climate change is measured as "global warming"—the increase in the average annual global temperature measured at the top of the atmosphere. Other measures of climate change (e.g., cloudiness, changing precipitation) could mean other shares for aviation than the ones cited here.

[16] Calculated by CRS from IEA data: "CO_2 Highlights 1971-2009." Paris, France, 2011.

[17] CRS calculated the estimate of CO_2 emissions from fossil fuels using IEA data, 1971-2009.

[18] ICAO statistics.

[19] Bureau of Transportation Statistics. "National Transportation Statistics." US Department of Transportation, Table 4-20. Data extracted January 23, 2012.

[20] Nancy N. Young, Vice President, Environmental Affairs, Air Transport Association of America, Inc., 2011 ACI-NA International Aviation Issues Seminar: Significant (and Challenging) Environmental Issues, December 1, 2011. According to National Transportation Statistics, international fuel consumption rose during this period, while total and domestic fuel consumption declined (NTS, Table 4-8, Certificated Air Carrier Fuel Consumption and Travel).

[21] Federal Aviation Administration, FAA's NextGen Implementation Plan, March 2011.

[22] SESAR Joint Undertaking, European Air Traffic Management Master Plan, Edition 1, Brussels, Belgium, March 30, 2009.

[23] Boeing Commercial Airplanes, Backgrounder: Boeing 787 Dreamliner Being Designed for Environmental Performance, Seattle, WA.

[24] Airbus, A350 XWB, available at http://www.airbus.com/aircraftfamilies/ passengeraircraft/ a350xwbfamily/.

[25] See, e.g., Max Kinsley-Jones, "No A320 Replacement 'Until 2024': Airbus," Flightglobal, available at http://www.flightglobal.com/news/articles/no-a320-replacement-until-2024-airbus-332442/.

[26] Geiver, Luke. "'If you build it, we will buy it' says airline CEO on biojet fuel," Biorefining Magazine. November 10, 2011. http://www.biorefiningmagazine.com/articles/5938/if-you-build-it-we-will-buy-it-says-airline-ceo-on-biojet-fuel.

[27] Airlines International. "Environment - The Growth of Biofuels." International Air Transport Association (IATA), January 2010. http://www.iata.org/pressroom/airlines-international/ december-2010/pages/environment.aspx.

[28] Hileman, James I., David S. Ortiz, James T. Bartis, Hsin Min Wong, Pearl E. Donohoo, Malcolm A. Weiss, and Ian A. Waitz. Near-Term Feasibility of Alternative Jet Fuels. Santa Monica CA: RAND Corporation, sponsored by the FAA, 2009. http://web.mit.edu/aeroastro/partner/reports/proj17/altfuelfeasrpt.pdf.

[29] ICAO. ICAO Environment Report 2010: Aviation and Climate Change. Quebec, Canada: published under the authority of the Secretary General, 2010. p. 31.

[30] Ibid. p. 33, and Figure 2 p. 32.

[31] Runge-Metzger, Artur. "Aviation and Emissions Trading: ICAO Council Briefing" presented at the ICAO Council, September 29, 2011. http://ec.europa.eu/clima/ policies/transport/aviation/docs/presentation_icao_en.pdf.

[32] For more information on the United Nations Framework Convention on Climate Change and on the Kyoto Protocol, see CRS Report R40001, A US-Centric Chronology of the International Climate Change Negotiations, by Jane A. Leggett.

[33] The EU Parliament and Council enacted this target in June 2009. The EU has also pledged that it would reduce its emissions to 30% below 1990 levels by 2020 if international agreement can be reached on a mandate to negotiate a legally binding treaty by 2015 that would include GHG mitigation obligations by all major emitting countries and that would take effect by 2020.

[34] For example, Decision No 280/2004/EC of the European Parliament and of the Council of 11 February 2004 concerning a mechanism for monitoring Community greenhouse gas emissions and for implementing the Kyoto Protocol; Directive 2003/87/EC of the European Parliament and of the Council of 13 October 2003 establishing a scheme for greenhouse gas emission allowance trading within the Community and amending Council Directive 96/61/EC; and 2010/778/EU: Commission Decision of 15 December 2010 amending Decision 2006/944/EC determining the respective emission levels allocated to the Community and each of its Member States under the Kyoto Protocol pursuant to Council Decision 2002/358/EC.

[35] Up to 1997, Parties disagreed about whether each Party should be responsible for emissions from sales of aviation fuels to international carriers or flights departing that country. The agreed GHG inventory methods required Parties to account for emissions from their sales of international bunker fuels but report them distinctly from the rest of the national total, pending policy decision. The Kyoto Protocol's Article 2.2 represents Parties' implicit understanding that aviation emissions would not be counted in each Party's "assigned amount" or cap on emissions.

[36] For a simple explanation of how an emissions cap-and-trade system works, see the Policy Tools section of CRS Report R41973, Climate Change: Conceptual Approaches and Policy Tools, by Jane A. Leggett.

[37] The ETS now includes emissions of nitrous oxides from some sources as well. For simplicity and because the nitrous oxide requirements do not apply to aviation, this memorandum will focus on CO_2 alone.

[38] http://www.sesarju.eu/about/background.http://www.sesarju.eu/about/background.

[39] EU. "Argumentaire: Inclusion of aviation in the EU ETS." Provided by the European Union Delegation to the United States of America, 2011.

[40] Directive 2008/101/EC of the European Parliament and of the Council of 19 November 2008 amending Directive 2003/87/EC so as to include aviation activities in the scheme for greenhouse gas emission allowance trading within the Community.

[41] De minimis criteria exclude from the ETS directive the following types of flights:

- those performed by aircraft with a maximum take-off weight of less than 5,700kg (roughly 12,500 pounds);
- those by operators with fewer than 243 flights for three consecutive four-month periods, or flights with total annual emissions lower than 10,000 metric tons per year.

[42] The threat of a ban is found in paragraph 5 of the amended Directive: "In the event that an aircraft operator fails to comply with the requirements of this Directive and where other enforcement measures have failed to ensure compliance, its administering Member State may request the Commission to decide on the imposition of an operating ban on the aircraft operator concerned."

[43] Directive 2008/101/EC, Article 3(d)(4).

[44] Currency exchange rates used in estimates differ among analyses and over time as well, contributing to as much as 20% of differences.

[45] The ATA has changed its name to Airlines for America, A4A.

[46] Airlines for America, "ATA Predicts US Industry, Economy Will Lose Billions due to Unlawful EU ETS," Washington, DC, July 27, 2011, available at http://www.airlines.org/Pages/ATA-Predicts-US-Industry,-EconomyWill-Lose-Billions-due-to-Unlawful-EU-ETS.aspx.

[47] Zisman, Matthew, and Randoph Walerius. EU Aviation Emissions: Impact on US Airlines and Freight Carriers. Bloomberg Government, November 29, 2011.

[48] Ibid.

[49] If the downturn in demand were sufficiently strong, operators could receive more allowances than they need, which could be "banked" for future use or sale when allowance prices rise. Some critics contend that giving aircraft operators a high percentage of allowances for free increases the potential for windfall profits for the sector. On the other hand, an economic downturn sufficient to render purchase of allowances unnecessary could have a severe impact on the sector.

[50] Maher, Sid. "Europe's $287bn carbon 'waste': UBS report." The Australian, November 23, 2011. http://www.theaustralian.com.au/national-affairs/europes-287bn-carbon-waste-ubs-report/story-fn59niix1226203068972.

[51] Ibid.

[52] Questions & Answers on the inclusion of aviation in the EU's Emission Trading System (EU ETS) (20/01/2012), http://eeas.europa.eu/delegations/russia/press_corner/all_news/ news/2012/20120120_en.htm.

[53] http://www.bloomberg.com/news/2011-09-26/airlines-may-use-carbon-permit-windfall-to-replace-aircraft-eusays.html.

[54] Malina, Robert, Dominic McConnachie, Niven Winchester, Christoph Wollersheim, Sergey Paltsev, and Ian A. Waitz. "The impact of the European Union Emissions Trading Scheme on US aviation." Journal of Air Transport Management 19, no. 0 (March 2012): 36-41.

[55] Nancy Trejos, "United-Continental, US Airways, Delta Add Europe Surcharge," USA Today, January 5, 2011.

[56] http://europa.eu/rapid/pressReleasesAction.do?reference=IP/11/1077. In other words, if the airlines charge passengers for the value of 100% of allowances (including the 82%-85% they get free), rather than the smaller percentage they may need to purchase to cover their emissions, consumers will pay more.

[57] Note, however, that these concerns reflect, according to economic theory, how market competition should motivate businesses to respond to the policy incentives, to innovate, and to adjust their emissions and costs. Those operators that are able to respond at least cost should gain market share relative to competitors that cannot.

[58] When fuel prices increase, they will tend to suppress demand for fossil fuels and resultant CO_2 emissions. Thus rising fuel prices will tend to drive down emission allowance prices. Conversely, when emission allowance prices are high, they will tend to suppress demand for fossil fuels and their prices. While large increases in fuel prices may tend to significantly moderate the cost of carbon allowances, increases in the cost of carbon allowances would likely have a much smaller influence on the effect of jet fuel prices and airline operating costs because they are likely to make up a comparatively small proportion of the total costs (as opposed to fuel prices which are a major component of airline costs).

[59] IATA, IATA Economics, Jet Fuel Price Monitor, price extracted 5 March 2012. http://www.iata.org/whatwedo/ economics/fuel_monitor/Pages/index.aspx.

[60] InterVISTAS Consulting, Inc. Estimating Air Travel Demand Elasticities: Final Report. Prepared for the International Air Travel Association, December 28, 2007. This report cautions that price elasticities of demand vary in important ways depending on the type of customer, the scope of price increases, etc. However, it states that the price elasticity of demand for trans-national price increases (i.e., an increase in price for all routes from a location) is about - 0.6. That is, for a 1% increase in price, demand might decrease by about 0.6%. Hypothetically, if the cost of a roundtrip transatlantic ticket increased by $6, or very roughly 1% of a discount economy fare, demand might decrease by about 0.6% (more among low-fare passengers than among high-fare passengers, who also tend to be less price sensitive).

[61] Ernst & Young and York Aviation. Inclusion of Aviation in the EU ETS: Cases for Carbon Leakage, October 31, 2008.

[62] 2008/101/EC, paragraph (9), http://eur-lex.europa.eu/LexUriServ/LexUriServ.do?uri= OJ:L:2009:008:0003:0021:en:PDF. The relevant text reads: At the sixth meeting of the ICAO Committee on Aviation Environmental Protection in 2004, it was agreed that an aviation-specific emissions trading system based on a new legal instrument under ICAO auspices seemed sufficiently unattractive that it should not then be pursued further. Consequently, Resolution A35-5 of the ICAO's 35th Assembly held in September 2004 did not propose a new legal instrument but instead endorsed open emissions trading and the possibility for States to incorporate emissions from international aviation into their emissions trading schemes. Appendix L to Resolution A36-22 of the ICAO's 36th Assembly held in September 2007 urges Contracting States not to implement an emissions trading system on other Contracting States' aircraft operators except on the basis of mutual agreement between those States. Recalling that the Chicago Convention recognises expressly the right of each Contracting Party to apply on a nondiscriminatory basis its own air laws and regulations to the aircraft of all States, the Member States of the European Community and fifteen other European States placed a reservation on this resolution and reserved the right under the Chicago Convention to enact and apply market-based measures on a non-discriminatory basis to all aircraft operators of all States providing services to, from or within their territory.

[63] This section was prepared by Daniel T. Shedd, CRS Legislative Attorney, 7-8441.

[64] Case C-366/10, Air Trans. Assoc. of America v. Sec. of State for Energy and Climate Change, 2011 available at http://curia.europa.eu/juris/document/document.jsf?text =&docid= 117193&pageIndex=0&doclang=EN&mode=lst& dir=&occ=first&part=1&cid=381567.

[65] Id.

[66] Id.

[67] The ATA has since changed its name to Airlines for America, A4A.

[68] The US-EU Air Transport Agreement, signed April 30, 2007. http://www.state.gov /e/eb/ rls/othr/ata/e/eu/ 114768.htm.

[69] Id.

[70] Id.

[71] Opinion of Advocate General Kokott, Case C-366/10, Air Trans. Assoc. of America v. Sec. of State for Energy and Climate Change, 2011 available at http://curia.europa.eu/juris/ document/document.jsf?text=&docid=110742& pageIndex=0&doclang=EN&mode=lst&dir=&occ=first&part=1&cid=381567.

[72] Id. at ¶ 52.

[73] Id. at ¶ 60, 71-72. The court noted that if the EU had assumed all of the powers exercised by the EU member States under the Chicago Convention, then the EU would be bound by the Convention. However, the court stated that because the powers "have not to date been assumed in their entirety by the European Union, the [European Union] is not bound by that convention." Id.

[74] Opinion of Advocate General Kokott, Case C-366/10, Air Trans. Assoc. of America v. Sec. of State for Energy and Climate Change, ¶ 71, 2011; Case C-366/10, Air Trans. Assoc. of America v. Sec. of State for Energy and Climate Change, ¶ 74, 2011.

[75] Id. at ¶ 74, 2011.

[76] Case C-366/10, Air Trans. Assoc. of America v. Sec. of State for Energy and Climate Change, ¶ 77-78, 2011.

[77] Id. at ¶ 111, 2011.

[78] Id.

[79] Id. at ¶ 131.

[80] Id. at ¶ 132-35.

[81] Id. at ¶ 121-30.

[82] Id. at ¶ 121-129.

[83] Id. at ¶ 136.

[84] Id. at ¶ 147.

[85] Id. at ¶ 148.

[86] Id. at ¶ 149.

[87] Id. at ¶ 150-51.

[88] Id. at ¶ 153-54.

[89] Id. at ¶ 155-57.

[90] Madhu Unnikrishnan, "US Threatens Action Against EU ETS," Aviation Daily, December 20, 2011.

[91] "US Demands Data From European and US Airlines As It Fires Off First Retaliatory Salvo In Growing Dispute With EU Over ETS," GreenAir Online.com, available at http://www. greenaironline.com/news.php?viewStory=1412.

[92] Airlines for America, A4A Comment on DOT Request for Airline Emissions Data, Washington, DC, December 16, 2011, available at http://www.airlines.org/Pages/news_12-16-2011.aspx.

[93] "Aviation and Emissions Trading," July 10, 2008, EurActive.com at http://www. euractiv. com/en/climate-change/ aviation-emissions-trading/article-139728.

[94] Regulations are sometimes characterized as "technology-following" or "technology-forcing." Technology-following regulations inscribe the performance of technologies already widely commercially deployed. By the time such regulations are adopted, most of the regulated entities will already exceed the standard. Technology-following standards act on so-called technology late-adopters or laggards. In contrast, technology-forcing regulations require, or

are based on, performance of leading-edge technologies that may be commercially in use by early adopters, or may be demonstrated but not yet commercially deployed on a significant scale. (In the latter case, the regulations allow a transition period from current to forced technology performance.) Most industrialized countries, including the United States, tend to rely on technology-forcing regulations. Some companies and countries use technology-forcing regulations to gain competitive advantage. ICAO's recommended standards are technology-following. See, for example, the discussion in Warwick, Graham. "Aircraft Emissions Standards Hard To Craft." Aviation Week (August 15, 2011). http://www.aviationweek.com/aw/jsp_includes/articlePrint.jsp?headLine= Aircraft%20Emissions%20Standards%20Hard%20To%20Craft&storyID=news/awst/2011/08/15/AW_08_15_2011_p45-352574.xml.

95 Mukherjee, Krittivas. "India, EU try to break logjam over airline carbon tax." Reuters Online, February 3, 2012. http://www.reuters.com/article/2012/02/03/uk-india-eu-airlines-idUSLNE81202520120203.

96 Flieger, Arthur. "China's Civil Aviation to Cut Emissions by 22% by 2020." Flieger Law, 2011. http://fliegerlaw.com/blog/?p=71. See also Smets, Isabelle. "Chinese airlines banned from participating in ETS." Europolitics. Brussels, February 6, 2012. http://www.europolitics.info/europolitics/chinese-airlines-banned-fromparticipating-in-ets-art325264-46.html: "Concerning China in particular, the Commission says it is prepared to discuss the possibility for airlines to be exempted from the system if China sets up 'equivalent measures' designed to reduce CO_2 emissions from air transport (this is a case for exemption foreseen in the EU Directive). Climate Action Commissioner Connie Hedegaard described as a 'step in the right direction' the publication by China's civil aviation authority, on 13 April 2011, of guidelines requiring national airlines to improve their energy efficiency and reduce their emissions by 22% from 2005 levels."

97 The Hindu Business Line. "Retaliatory Action Likely Against EU for Carbon Tax on Airlines from India." Jan. 30, 2012. http://www.thehindubusinessline.com/industry-and-economy/article2845392.ece.

98 "Joint declaration of the Moscow meeting on inclusion of international civil aviation in the EU-ETS." As reported by Russian Aviation. Moscow, 22 February. http://www. ruaviation.com/docs/1/2012/2/22/50/print/.

99 "UPDATE 4-Foes of EU airline CO_2 rules agree on tactics." Reuters.com, February 22, 2012. http://www.reuters.com/article/2012/02/22/eu-airlines-idUSL5E8DM3YD20120222.

100 EU Ambassador to New Delhi Joao Cravinho, as quoted in Euroactiv.com, "EU envoy pushes airline carbon law waivers," 24 January 2012. http://www.euractiv.com/climate-environment/eu-envoy-pushes-airline-carbon-lawwaivers-news-510351.

101 Article 2.2 of the Kyoto Protocol.

102 ICAO, "Statement from the International Civil Aviation Organization (ICAO) to the thirty-fifth Session of the UNFCCC Subsidiary Body for Scientific and Technological Advice (SBSTA 35)" (Durban, South Africa; November 28 to December 3, 2011).

103 Ibid.

104 The first allocation of emission allowances to aircraft operators is February 28, 2012.

105 See, for example, Bartels, Lorand. "The Inclusion of Aviation in the EU ETS: WTO Law Considerations." ICTSD Trade and Sustainable Energy Series No. 6 SSRN eLibrary (November 15, 2011) (suggesting that the EU ETS may not be justifiable under WTO law).

106 IACA, "Emissions Trading Concerns Will be Solved Through ICAO, not in Court." News and Views, 1 January 2012, http://www.iaca.be/index.cfm?9AAC75F4-1851-F050-B2AF-25B0B4948D39.

[107] http://unfccc.int/2860.php.

[108] Chicago Convention Article 37.

[109] Chicago Convention Articles 11 and 15.

[110] The United States is a member of CAEP.

[111] The United States is on the ICAO Council.

[112] ICAO. Report of the Executive Committee on Agenda Item 15. ICAO, December 4, 2004. ICAO.

[113] ICAO makes a conceptual distinction between taxes—with the purpose of raising revenues—and levies—with the purposes of cost recovery for services provided. In the case of environmental levies, they would apply to use of the atmosphere or some other publicly held resource for disposal of private waste (including engine emissions).

[114] An "open" trading system is one in which allowances allocated to or required in one sector may be traded with allowances from other covered sectors. In other words, and aircraft operator would be able to sell allowances to brokers or facilities in other sectors, or to purchase them in the multi-sectoral market. In a closed trading system, allowances could be bought and sold only within the same sector. Aviation in the EU ETS is an open trading system.

[115] ICAO. 2004. op. cit., Appendix I.

[116] ICAO. "A37-19: Consolidated statement of continuing ICAO policies and practices related to environmental protection—Climate Change." October 2011. http://www.icao.int/sustainability/Documents/A37-Env-Resos-9958.pdf. Clauses 4 and 5.

[117] Ibid. Clause 6.

[118] For more information on assistance to low income countries to address climate change, see CRS Report R41808, International Climate Change Financing: Needs, Sources, and Delivery Methods, by Richard K. Lattanzio and Jane A. Leggett.

[119] "Capacity building" is a term of art often used to describe assist to developing countries to enhance governance and the effectiveness of the private sector. It may cover assistance to improve the rule of law, civil engagement, efficiency of administration, training for private sector activities, etc. Some capacity building is provided through public channels and some through private and other non-governmental channels.

[120] ICAO Assembly Resolution A33-7 Appendix I, adopted by the 33rd Session of the ICAO Assembly in September/October 2001.

[121] The "technology following" approach is in contrast to a "technology forcing" approach, which sets standards based on technologies that are demonstrated and commercially available, but in use only by the leading edge (a minority) of manufacturers and operators.

INDEX

A

B

C

D

E

F

G

V

W

Y